THE VISIBLE HUMAN PROJECT

The Visible Human Project is a critical investigation of the spectacular, three-dimensional recordings of real human bodies – dissected, photographed and converted into visual data files – made by the US National Library of Medicine in Bethesda. Catherine Waldby uses new ideas from cultural studies, science studies and social studies of the computer to situate the Visible Human Project in its historical and cultural context, and to consider the meanings such an object has within a computerised culture.

In this fascinating and important book, Catherine Waldby explores how advances in medical technologies have changed the way we view and study the human body, and places the VHP within the history of technologies such as the X-ray and CT scan, which allow us to view the human interior.

Bringing together medical conceptions of the human body with theories of visual culture from Foucault to Donna Haraway, Waldby links the VHP to a range of other biomedical projects, such as the Human Genome Project and cloning, which approach living bodies as data. She argues that the VHP is an example of the increasingly blurred distinction between 'living' and 'dead' human bodies, as the bodies it uses are digitally preserved as a resource for living bodies, and considers how computer-based biotechnologies affect both medical and non-medical meanings of the body's life and death, its location and its limits.

Catherine Waldby is Senior Lecturer in Media, Communication and Culture at Murdoch University and Deputy Director of the National Centre in HIV Social Research, University of New South Wales, Australia.

BIOFUTURES, BIOCULTURES
Series Editor: Catherine Waldby

THE VISIBLE HUMAN PROJECT

Informatic bodies and posthuman medicine

Catherine Waldby

London and New York

First published 2000
by Routledge
11 New Fetter Lane, London EC4P 4EE

Simultaneously published in the USA and Canada
by Routledge
29 West 35th Street, New York, NY 10001

Routledge is an imprint of the Taylor & Francis Group

© 2000 Catherine Waldby

Typeset in Garamond by Taylor & Francis Books Ltd
Printed and bound in Great Britain by Biddles Ltd, Guildford and King's Lynn

All rights reserved. No part of this book may be reprinted or reproduced or utilised in any form or by any electronic, mechanical, or other means, now known or hereafter invented, including photocopying and recording, or in any information storage or retrieval system, without permission in writing from the publishers.

British Library Cataloguing in Publication Data
A catalogue record for this book is available from the British Library

Library of Congress Cataloging in Publication Data
Waldby, Catherine
The Visible Human Project : informatic bodies and posthuman medicine / Catherine Waldby.
Includes bibliographical references and index.
1. Visible Human Project. 2. Human anatomy–Atlases. 3. Human anatomy–Databases. 4. Medical informatics. 5. Body, Human (Philosophy) I. Title.
[DNLM: 1. Visible Human Project. 2. Anatomy, Cross-Sectional. 3. Bioethics. 4. Image Processing, Computer-Assisted. 5. Models, Anatomic.
QS 4 W157v2000]
QM25 .W25 2000
611'.0022'2–dc21   99-088733

ISBN 0–415–17405–8 (hb)
ISBN 0–415–17406–6 (pb)

CONTENTS

List of figures vii
Acknowledgments ix
List of acronyms xii

1 The Visible Human Project: an initial history 1

2 Posthuman spectacle 24

3 Theatres of violence: the anatomical sacrifice and the anatomical trace 51

4 Virtual surgery: morphing and morphology 81

5 IatroGenesis: digital Eden and the reproduction of life 110

6 Revenants: death and the digital uncanny 136

7 Technogenesis: the posthuman visible 157

Notes 163
Bibliography 170
Index 179

FIGURES

1.1	Three-dimensional reconstruction of the male dataset: head	2
1.2	Three-dimensional reconstruction of the female dataset: full body	3
1.3	CT scan of Visible Woman	10
1.4	MRI of Visible Male head	11
1.5	Section through Visible Male: thorax	14
1.6	Section through Visible Male: skull	15
1.7	Flythrough of Visible Male bone surfaces	16
3.1	Plates showing some of the muscles, from Andreas Vesalius' *De Humani Corporis Fabrica*, 1543	65
3.2	Dissection showing the hyperglossal and lingual nerves, in *Gray's Anatomy*	68
3.3	Heart model computed from Visible Male data	75
3.4	Visible Male Head showing interior structures	76
3.5	CT scan of Visible Male thorax	77
4.1	CT reconstruction showing infant skull	83
4.2	Three-dimensional reconstruction of knee using VHP data	85
4.3	Visible Male pelvis with prostate gland from cancer patient	87
4.4	Three-dimensional reconstruction of the Visible Human Male, showing skin, internal organs and skeletal structure	97
4.5	Heart model computed from Visible Male dataset, and three virtual endoscopic views showing interior surfaces of major vessels and heart chambers	104
5.1	Adam and Eve figures, in *Compendiosa Totius Anatomie Delineatio* by Thomas Geminus, 1545	112
6.1	Melancholy skeleton, in *Compendiosa Totius Anatomie Delineatio* by Thomas Geminus, 1545	140

ACKNOWLEDGEMENTS

The idea for this book began in a café. Sitting out on the pavement in the blinding Fremantle sunshine a friend said, 'Did you hear about the guy that was cut up and put on the web?' I hadn't heard, and over the next few days surfed around getting more information. I found the strange, poignant images of Jernigan (the Visible Woman had not yet been launched) utterly compelling, so compelling that they propelled me into writing this book. In a sense this book is my way of understanding why I found the images so fascinating and why they seemed to exercise the same attraction for others. In answering this question I have had the opportunity to examine the uncanny productions of biomedicine and biotechnologies more generally, an examination which took this text into some decidedly gothic, posthuman realms. I hope that my interpretation of the Visible Human Project resonates with my readers' own responses to these images, and helps to account for their disturbing power.

As with any academic text, the work involved in turning an original idea into a final product is enormous. I owe numerous debts to particular people and institutions for their assistance in carrying out this work.

First of all I would like to thank my home institution, the School of Media, Communication and Culture at Murdoch University. The School, and the University more generally, have extended numerous kinds of support to this research. It has received several research grants, including two ARC small grants and three other rounds of grant money, which enabled me to travel to several VHP sites around the world. I would particularly like to thank my program chair Alec McHoul for his unfailing support and encouragement, both institutional and personal, and the school research support officer Peter Stuart for his wonderful computer skills and enthusiastic help. I would also like to thank the Department of Sociology, University of N.S.W. in Sydney, which awarded me an honorary sabbatical fellowship during 1998, when much of the final drafting of this book was prepared. In particular I would like to thank Anne Daniel for her support.

Second I would like to thank the many scientists at a variety of VHP sites throughout the world who generously extended their time, ideas and assistance to me, despite the fact that I am a cultural studies academic rather than a fellow scientist. In particular I wish to thank: Brian Molinari, Paul Mackerras and David Hawking at the Department of Computer Sciences at the Australian National University, Canberra, Australia; the staff at the Institute for Mathematics and Computer Science in Medicine at the University Hospital Eppendorf, Hamburg, Germany. I would particularly like to thank Thomas Schiemann for his assistance; the staff at the Centre for Information-Enhanced Medicine, National University of Singapore, particularly Dr Mullick and Dr Raghavan; Michael Ackerman, VHP Project Officer, National Library of Medicine, National Institutes of Health, USA; and Victor Spitzer, Director of the Centre for Human Simulation, University of Colorado, USA.

Third I would like to thank Celia Roberts, Wendy Parkins, Ingrid Richardson and Suzanne Fraser for various kinds of research assistance. As anyone who has ever worked as a research assistant knows, the work varies from the most banal tasks to the highest level of intellectual collegiality. I am grateful to each of these women for their insights as much as for their more practical contributions. I am particularly grateful to Ingrid Richardson for more kinds of practical and intellectual assistance than I can describe here.

Fourth I am indebted to the following individuals and institutions for various kinds of ideas, support and assistance: Elizabeth Fee and the other staff at the History of Medicine Library, National Library of Medicine for their encouragement and assistance; Lisa Cartwright for her unfailing collegiality and intellectual generosity; Nancy Condee, Jane Feuer, Kathy Ferraro, Brad Lewis, Carole Stabile and the rest of the University of Pittsburgh crew for their wonderful hospitality and encouragement; Akira Lippit, Janine Marchessault, Jutta Doberstein, Fred Truniger and Herbert Schwarze for the inspired delirium of the Oberhausen Film Festival; my graduate students Nikki Miller, Barbara Bolt and Carly Harper for many discussions about the image, aesthetics and medicine; the members of the Science Studies reading group at Sydney University, particularly Adrian Mackenzie, Paul Griffiths and Anne-Marie Johnson, for their incisive discussions of bioinformatics; Elizabeth Wilson for her friendship and her passionate, critical engagements with science; Cathryn Vasseleu for her brilliant insights into virtual light; Moira Gatens, Sean Cubbitt and N. Katherine Hayles for their kind and useful comments on earlier versions of this manuscript; and all those other friends, students and colleagues, too numerous to name, who have discussed aspects of this work with me.

Finally I would like to thank my partner Mark Berger, who had to live through the writing of this book day by day, and put up with my many authorial neuroses. Without his patience, love and financial support the work would have been infinitely slower and more painful.

Earlier versions of some chapters and ideas in this book appeared in the following publications: 'The Visible Human Project: Data into Flesh, Flesh into Data', in K. Sawchuck and J. Marchessault (eds) (2000) *Wild Science: Feminist Readings of Science, Medicine and the Media*, London: Routledge; 'IatroGenesis: the Visible Human Project and the Reproduction of Life', *Australian Feminist Studies* Special Issue on Feminism and Science, vol. 14, no. 29, 1999: 77–90; 'Virtual Anatomy: from the Body in the Text to the Body on the Screen', *The Journal of Medical Humanities* vol. 21, no. 1, 2000; 'The Body in the Digital Archive: the Visible Human Project and the Computerisation of Medicine', *Health: an Interdisciplinary Journal for the Social Study of Health, Illness and Medicine* vol. 1, no. 2, 1997: 227–243; 'Revenants: the Visible Human Project and the Digital Uncanny', *Body and Society* vol. 3, no. 1, 1997: 1–16; 'The Visible Human Project and the Digital Uncanny', *44th Internationale Kurtzfilmtage Oberhausen Catalogue*, Oberhausen, Germany, 23–28 April 1998, pp 121–125.

IMAGE ACKNOWLEDGEMENTS

The author would like to thank the following people and institutions for permission to use their images: William T. Katz, Varian Medical Systems; The National Library of Medicine, National Institutes of Health; The Institute of Mathematics and Computer Science in Medicine, University Hospital Eppendorf, Hamburg; Muriel D. Ross at the NASA Ames Research Center; Richard Robb, Ph.D., Biomedical Imaging Resource, Mayo Foundation; Bill Lorensen at the GE Imaging and Visualization Laboratory.

ACRONYMS

AIDS	Acquired Immunodeficiency Syndrome
CAD	Computer-Aided Design
CMC	Computer Mediated Communication
CT	Computed Tomography
DNA	Deoxyribonucleic Acid
HGP	Human Genome Project
HIV	Human Immunodeficiency Virus
IVF	In-Vitro Fertilisation
MRI	Magnetic Resonance Imaging
NIH	National Institutes of Health (Bethesda, Maryland, USA)
NLM	National Library of Medicine (Bethesda, Maryland, USA)
PET	Positron Emission Tomography
VHD	Visible Human Data
VHP/VH	Visible Human Project, Visible Human
VM	Visible Man
VW/VF	Visible Woman, Visible Female
VR	Virtual Reality

1
THE VISIBLE HUMAN PROJECT
An initial history

In a *fin-de-millennium* culture well accustomed to spectacle and dramatic biotechnical innovation, the public launch of the Visible Human Project created an extraordinary global *frisson*. When the National Library of Medicine, the authors of the project, made the first images available on their website in November 1994 they immediately caught the attention of the popular media as well as the scientific community.[1] Throughout the world a fever of TV documentaries, press articles and radio programmes ensued. The project's iconography has since featured in every mass circulation magazine from *Time* to *Wired*, as well as the more soberly scientific publications. In general the articles and programmes I remember or have since read or watched were quite straightforward. They described the process of the VHP figures' production (a gruesome tale in itself and one which I will soon retell here), provided a selection of images and viewpoints of the virtual body, and quoted an assortment of scientists speaking on the possible uses for the project. Most media coverage gave extensive attention to the first human to become a Visible Human, one Joseph Paul Jernigan, of Waco, Texas. Jernigan had been on death row in a Texas prison for twelve years, convicted in 1981 for burglary and murder. In August 1993 Jernigan was executed by injection with a lethal dose of potassium chloride. The choice of Jernigan's body by the project team provided media coverage with a set of stock narratives and an appealing moral economy of criminal transgression, punishment, sacrifice and redemption which produced headlines like 'Executed man helps science as internet cadaver', 'Executed killer reborn as visible man on internet' and 'A convict's contribution'.[2] (See Figure 1.1.)

Late in 1995 a second body was launched for the project, a Visible Woman, based on the body of an unnamed 59-year-old woman, described in the press as a 'Maryland Housewife' who died of a heart attack, and whose body is said to have been donated to the project at the explicit request of her husband. Predictably enough this development was immediately interpreted by the media as the creation of a kind of virtual couple, and perhaps the beginnings of a virtual family in cyberspace. (See Figure 1.2.)

THE VISIBLE HUMAN PROJECT

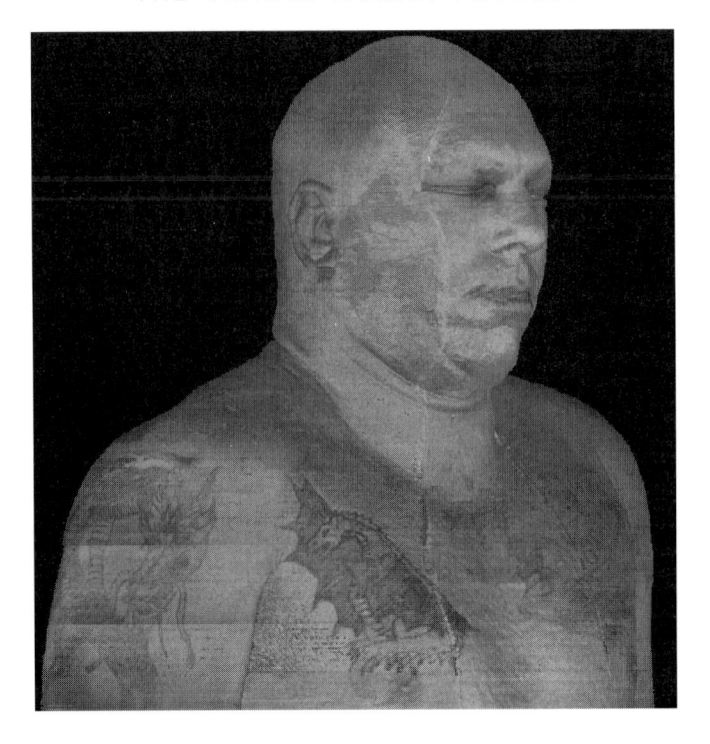

Figure 1.1 Three-dimensional reconstruction of the male dataset: head (image courtesy of William Katz, Varian Medical Systems)

The relative banality of these media stories seems incommensurate with the sheer volume of attention the project received and still receives in popular and public culture, an attention primarily addressed to the project's dramatic visuality. A cursory browse of the World Wide Web will produce literally thousands of sites dedicated to or linked to the project, where samples of the cross-sections, reformulations and animations can be viewed. The VHP figures have been the centrepieces for several major gallery or museum exhibitions. In 1995–6 a gallery exhibition in Japan juxtaposed images from the project with Da Vinci's anatomical drawings. The Smithsonian Institute has dedicated a station to the project. The project formed the subject of a major exhibition which was first at the Maryland Science Center in Baltimore in 1998 and then travelled throughout the USA. A French film, *The Fifth Element*, released in 1997, used the data as the basis for a special-effect sequence. The scene, itself an homage to *Metropolis*, uses a sequence of the VHP tissue cross-sections to depict the creation of an improbably seductive female humanoid in a laboratory. In 1997 it was the subject of an extensive visual spread in a dedicated issue of *Life* magazine, 'A fantastic voyage through the human body', featuring the highly aestheticised

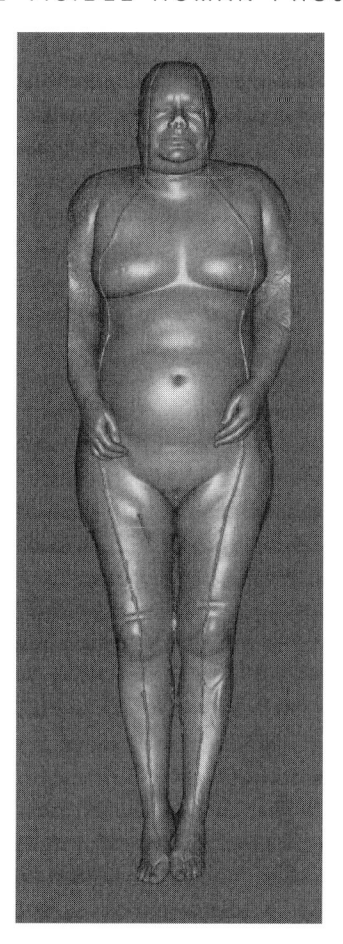

Figure 1.2 Three-dimensional reconstruction of the female dataset: full body (image courtesy of B. Lorensen, G.E. Imaging and Visualisation Laboratory)

work of medical photographer Alexander Tsiaras, which also compared the Project images to Da Vinci's anatomies, and to Rembrandt's 'The Anatomy Lesson of Dr. Nicolaes Tulp'.[3] Intel Computers used the data for an advertising campaign during 1997 that compared the inside of the computer 'brain' with the interior of a human brain. The licensees include Hollywood production companies, and Michael Ackerman, the project's CEO, predicts that it is only a matter of time before the data are reworked as special effects for mainstream US cinema. The virtual corpses figured in the VHP have become objects of medical and popular globalisation, circulating through the net, the cinema, and other visual mass media, an index of the interpenetration of popular culture with science.

It seems that the fascination the project compels, and I include my own fascination here, is widely acknowledged but difficult to articulate, a fascination with the spectacularly gruesome iconography of the project, which retreats from the task of critical analysis. For the most part, the pitch of media attention has simply taken its cue from the progressivist tone which biomedical scientists habitually use to frame their innovations. The majority of articles has presented anodyne, congratulatory assessments along the following lines:

> We can now access a spectacular head-to-toe study of the human body via the computer. Through the advent of computers ... students in the radiological sciences and other areas of medicine will now gain a better understanding of anatomy. Likewise, in the near future, children accessing computers will find out more about health, while physicians will find out more about cancer growth, radiation exposure levels and new ways to simulate surgery. All of this will be possible through the computerised study of the human body, called the Visible [Human Project].
>
> (Hartfield 1995)

The Visible Human Project is, in these terms, simply another step along the road of medical progress, and another tool in the advance of knowledge, albeit a visually arresting tool with rather science-fictional overtones. It provides, according to this way of thinking, a better window into a pre-given human physiology and anatomy, a more efficient and transparent way to communicate this anatomy to viewers, a better medium for a given content. Just as Da Vinci's ratiocinative anatomies have served as icons for humanist knowledge and technical modernity, so the Visible Human Project has been taken up as a new iconography of 'Man' for the virtual future, a future in which all content, even the mysterious materiality of the human body, can be hyper-mediated, transported and traversed by the computer. It circulates in the media as evidence of the near-future, token of empowering biotechnical transformations yet to come.

Yet, like all fascination, the compelling attraction of the VHP figures travels along the edge of sublime disturbance. Their first appearance seemed to fracture a quotidian sense of space, and focus a global public attention on the strange potentials of the virtual. As many commentators (e.g. Wark 1993; Robins 1996; Benedikt 1994) have pointed out, virtual space and cyberspace have unique and unprecedented imaginative possibilities, philosophical implications and ontological consequences that can only be realised and explored in particular practices. The possible effects of the existence and functions of this new space, and its relationship to everyday space, are open questions, and the subject of intense cultural examination. While the possibilities of virtual space and cyberspace had been, by the mid-1990s, well explored in popular culture, the sudden appearance of data 'recordings' of real human bodies and the digital 'preservation' of the dead realised themes that had seemed, until that point, purely speculative. Themes of doubling and iterability (the film *Multiplicity*, 1996), reanimation and criminal virtual ontologies (the film *Virtuosity*, 1995, the *Neuromancer* series of novels) and the digitisation of subjectivity (the film *Johnny Mnemonic*, 1995) had all been played out in those other virtual spaces, cinema and fiction.[4] The arrival of the Visible Human Project seems to confer factual realisation on what had been considered merely science fiction. In particular it enacts the proposition that the interface between virtual and actual space, the screen itself, is permeable, rather than an hygienic and absolute division. This proposition is in the process of being thematised in a number of sites of cyberculture, most notably in the realm of virtual reality, which projects the user into immersive, virtually generated environments. Equipped with data goggles and a data glove, the virtual reality cybernaut moves through data landscapes as if they were physically located within that space, able to navigate via visual and haptic feedback. Nevertheless the inauguration of the Visible Male suggests a new itinerary for this idea, the possibility that the virtual threshold could be definitively transgressed through a violent and posthumous process, a forced transgression with connotations of punishment and incarceration, rather than the masterful and reversible transgression enacted by the VR user. The VHP confirms a certain fear easily suggested by screen technologies; the fear of being caught on the far side of the screen, lost on the virtual side of the looking-glass.[5] The masterful user of computers, subject of knowledge, is confronted with the possibility of absorption into the computer and transformation into some weird digital ontology. Virtual space had in this moment became a strange mirror space, where 'one could see oneself, before oneself, transformed'.[6]

Equally disturbing and compelling, the visibility involved in the Visible Human Project exposes the bodily interior to new orders of machine vision and presents a relentless public visual access to every organ, every corporeal manifold, from multiple new points of view. In demonstrating this new form of corporeal transparency the VHP figures an experience of visual objectification which is becoming more and more common for citizens of the industrialised nations. New forms of medical imaging – magnetic resonance imaging (MRI), computed tomography scans (CT) and their many variations – involve the enclosure of the patient's body within a 'total optical system' (Kember 1991: 59), which reads the body's interior as digitised information configured on a computer screen.[7] The new perspectives opened on to the bodily interior by the VHP and these other modes of vision suggest that the body at the end of the second millennium is utterly available as visible matter. Such a development is simultaneously reassuring, promising more accurate diagnoses and surgical procedures, an ever-increasing ability to locate and excise frightening interior pathologies, and disturbing, suggesting a kind of vulnerability to technical knowledge and machinic vision that generates a certain anxiety. In Don DeLillo's novel *White Noise* the central character, Jack Gladney, refuses a second encounter with an MRI scan on grounds of its excess knowledge.

> Dr Chakravarty ... wants to insert me once more in the imaging block, where charged particles collide, high winds blow. But I am afraid of the imaging block. Afraid of its magnetic fields, its computerised nuclear pulse. Afraid of what it knows about me.
>
> (DeLillo 1985: 325)

In this sense the VHP figures and their cognate forms of diagnostic imaging, describe some new interiority, a projected space of private, psychic being made globally visible and available through the distributive optics of the computer or the internet. The abolition of the bodily interior as private, or sacred, space, a process begun with the earliest systematic anatomies of the late Medieval period and extended with the abrupt application of x-rays to the body in 1895, takes on a new vigour with the launch of Visible Human figures.

Each of these strange transformations takes on a further uncanny aura, magnified by the project's eschatological implications. As virtual apparitions of dead bodies, the VHP figures seem possessed of disconcerting presence and highly uncertain status. These corpses were dismembered, their flesh effectively destroyed in the process of imaging, yet they reappear as recomposed, intact and reanimated bodies in a virtual space, 'copied' into the eternal medium of data. They are consequently difficult to locate within any proper distinction between the living and the dead. The VHP confirms virtual space as a domain of paradoxical incarnations and ambiguous production, haunted by the immaterial embodiment of the ghost suspended in some weird digital afterlife, or trapped in the horror of reanimated death. If the figures prefigure some new future for the human body, they imply the possibility of frightening, rather than consoling, transformations. This new power of virtual technologies to 'copy' a body does not simply work as a benign reflection, a symmetrical moment in which the human form finds it confirming virtual analog. Rather, like all doubles, the figures displace what they seem to mirror, and like all apparitions they return the familiar in unfamiliar form – the human reproduced and destabilised by the Visible Human.

The transfiguration of a human body into digital substance brings a whole domain of biotechnology into play, all those recent and promised innovations which image or engineer human bodies as archives of data. The scope of such technologies is already vast: the new forms of medical imaging (positron emission tomography (PET), MRI, CT scans) which traverse and image the body through the digitisation of electromagnetic or radiation quanta; genetic engineering technologies which test or rewrite DNA information in the organism; cloning techniques which copy genomes. In particular the Visible Human Project presents a kind of 'sister' project to the Human Genome Project, a scheme to map and sequence every gene in 'the human body', a scheme which is, we are told, 'the key to what makes us human, what defines and limits our possibilities as members of the species Homo sapiens' (Kevles and Hood 1992: vii). The VHP also claims to be an exhaustive archive of human information, although unlike the tedious productions of the Human Genome Project (endless strings of nucleotide sequence – TGCCTGGACTT) its information is, literally, spectacular. Both these 'Human Projects' present themselves as new ways to map and know the human, yet both imply a disconcerting threat to any idea of the human as a stable, knowable 'species', an organic integrity whose limits can be positively specified. If human bodies can be rendered as compendia of data, information archives which can be stored, retrieved, networked, copied, transferred and rewritten, they become permeable to other orders of information, and liable to all the forms of circulation, dispersal, accumulation and transmission which characterise informational economies. Any fantasy of organic integrity is lost in the face of the interface, the potential for data bodies to be integrated into data circuits, cybernetic or genetic. The Visible Human Project, read against its claims to Human definition, seems to present a vision of such permeability taken to its logical conclusion. Here the body is not simply a component in an interface but is utterly absorbed into data and data space, a digital ghost in the machine.

It seems likely that, in the relentless turnover of popular culture, the strangeness of the Visible Human Project registered only briefly in the public imagination, a strangeness whose depths were never adequately sounded. If, as I claim, its popular reception fractured a certain everyday working of space, and upset a certain idea of human status, these disturbances are rapidly glossed over, reabsorbed, and forgotten as media attention moves elsewhere, to the next scientific breakthrough or scandal. My task in this study is to return to this moment of disturbance and subject the strangeness of the VHP to an extended critical interrogation, to articulate the fascination which its iconography exercises.[8] In summoning up this virtual apparition I hope to make it speak to a number of questions. Some of these questions relate to biopolitics. In particular the VHP raises the general biopolitical question of medical objectification, the tension between biomedicine as a practice which on the one hand augments and shores up the status of the human, protecting subjects from the encroachments of diseased embodiment, and on the other generates its knowledges and procedures by treating the human as experimental object and passive biomass. It also raises the issue of medicine's medium specificity, the extent to which its knowledge of bodies, and its abilities to work them, is conditioned by the medium of objectification, rather than through some direct encounter with the full presence of flesh. The VHP is, as I will discuss at length, an exemplary instance of biomedicine's relentless refiguration and textualisation of the human body according to instrumental logics of knowledge, the particular organisation of its tools, rather than according to a pre-existing biological organisation. To this extent it implicates an entire field of computer-based medical imaging, and this text will link the VHP with a range of other practices which address living bodies as visual data.

Other questions relate to the VHP as a kind of fantasy object for medicine, a form of pornography which plays out medicine's imaginary anatomies in unprecedented ways. Pornographic and medical genres frequently converge around the quest for a maximised bodily visibility and the performative visuality of the VHP lends it to extended biomedical reveries about scientifically co-operative forms of corporeality, the possibilities of a perfectly accessible and mastered form of embodiment. Still others relate to the gothic texture of biomedicine which lines the inside of its self-consciously anti-gothic, rational aesthetic. The VHP is only one of many uncanny figures and entities which emerge from medicine's avowed desire to subject the reproduction of bodies to rational, standardised control, and its general project to refigure the relations between the living and the dead, to make death residual. Medicine's use of data and data space is itself uncanny, drawing on the peculiarly vivid, negentropic qualities of information to (re)animate its productions.

In addressing these and other questions my intention is, above all, to use the VHP as a spectre which haunts the human with the fragility of its status as subject in a world of objects. The VHP, read one way, seems a testament to human mastery over the natural world, mastery demonstrated through the ability to technically synthesise a body homologue, to re-place nature. Read another way it throws open the human to multiple incursions, demonstrating the body's possibilities for commodification, for instrumentalisation, demonstrating also its use value within technically driven orders of rationality. Rather than the human as the origin of technology, the VHP suggests a technogenesis for the human, the openness of the category 'human' to technical production and reproduction. In what follows I want to use the VHP as a way into this fragility, the always failing distinction between inventor and invention which underpins biomedicine, and which must constantly be re-placed and renegotiated in the face of biotechnical developments.

THE VISIBLE HUMAN PROJECT: A BRIEF HISTORY

The history of the Visible Human Project is, in one sense, coextensive with the history of scientific medicine. As an anatomical image it is produced out of a set of problems – of visibility, of the rendering and management of the bodily interior, of relations between part and whole and between the normal and the pathological – which can be identified in the earliest moments of scientific anatomy. Nevertheless, here I will confine my account to its immediate history and context as a medical tool, as a first, pragmatic introduction to my topic. The continuities between the VHP and earlier anatomical paradoxes and problems are taken up at length elsewhere, particularly in Chapter 3. Many other points made in this initial account reappear for in-depth treatment and problematisation in subsequent chapters.

For the biomedical community the VHP is simply a new kind of visual tool, albeit a particularly complex and useful one, a new kind of window through which to see the human body. Or perhaps it would be more accurate to describe it as a new kind of body-image rendered amenable to computer vision, the 'window' that medicine increasingly deploys in contemporary practice. The immediate impetus for the project, first mooted in 1988, derived from a rapid expansion in the use of computerised vision for diagnostic imaging over the previous decade. Since the early 1970s new computerised techniques for imaging the body's interior had been gradually replacing or supplementing the traditional radiograph, a photographic technology. These techniques involve the digitalisation of forms of radiation passed through or emitted by the body, and then the rendering of this digital data as a visual representation of the body on the computer screen.

The most commonly used of these is computed tomography (CT scan) and magnetic resonance imaging (MRI). CT scans are based on the technique of the x-ray, but rather than providing a 'shot' from just one angle, as conventional radiography does, the scan uses a thin beam which rotates around the entire body and takes a visual 'slice' through a cross section of the body. The radiological information, instead of being developed on a photographic medium to produce an x-ray image, is translated into digital data which is then translated into visual data on a computer screen, displaying a colour coded and enhanced cross section of the body. (See Figure 1.3.)

MRI is an electromagnetic technique which realigns hydrogen atoms in the body, causing them to emit a small electric current. A computer translates this current into an image of the area scanned (Sochurek 1987). (See Figure 1.4.)

Other diagnostic technologies include digital subtraction angiography, radioisotope imaging, sonography, and positron emission tomography, all techniques which are useful for imaging certain domains of the body or certain conditions, and which rely on the computer's ability to render all forms of information as visual information.

In addition to these new ways to image body data, the computerisation of medical vision had, by the late 1980s, developed ways to render such data in virtual volume, three dimensionally. For example, the cross-sectional data generated in CT scans can be translated into a three-dimensional visual model of a patient's skull by restacking the cross-sections. The computer's ability to represent and manipulate volume was hence transforming the space of medical thought itself, presenting it with novel 'points of view' and

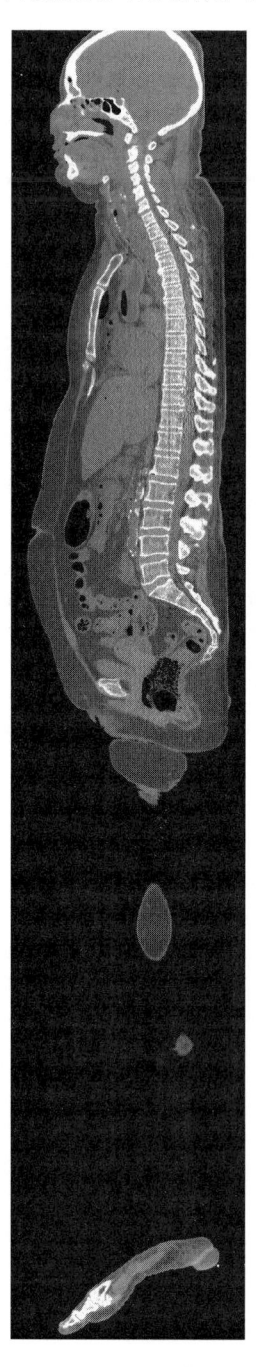

Figure 1.3 CT scan of Visible Woman (image courtesy of the National Library of Medicine, USA)

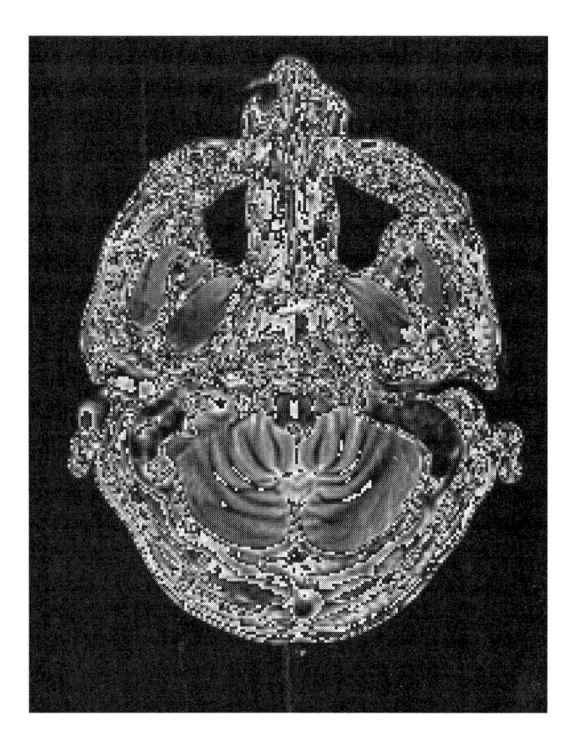

Figure 1.4 MRI of Visible Male head (image courtesy of the National Library of Medicine, USA)

new practices of modelling, surgical planning and visual pedagogy. In particular it presented a solution to the problem of moving from two-dimensional images of anatomy (x-rays, drawings) to three-dimensional bodies, a central problem in medical pedagogy as medical students learn the difficult task of envisaging the complex anatomical structure of the body volumetrically.

The necessity for a Visible Human archive was first posed in 1988 as a response to a perception of certain kinds of disorder and lack in the burgeoning field of computerised medical imaging. The National Library of Medicine (NLM), a division of the US National Institutes of Health (NIH), became an organising point for these concerns. In 1986 the National Library had given official recognition to the importance of digital imaging in the formation of medical knowledge, creating a committee to develop a digital imaging database. With the development and refinement of digital image transmission, the NLM recognised the possibility that anatomic images, including volumetric images, could be stored, retrieved and distributed in the same way as its text-based data, through networked computers (Masys 1990). The computer now presented the possibility that the NLM and other similar medical libraries could take on the role of a networked anatomical theatre, taking advantage of the distributive optics of the internet to create a new audience and usership for indexed anatomic iconography. In 1988 a consortium of US-based academic medical centres urged the NLM to give greater attention to the growing centrality of computer-based, three-dimensional imaging techniques in medical practice.

> Each of the university groups presented an overview of the types of activities at its own centre. In the aggregate it became clear that the power, graphic display capabilities, and affordability of current computers are sufficient for many educational and research applications in 3-D anatomical imaging, and that there are unique merits to images rendered from digitised anatomic databases. The most dramatic of these is the ability to isolate, highlight, 'reversibly dissect', rotate, and view from multiple angles single and grouped tissues, organs, body regions, and physiological systems.
>
> (Ackerman 1992: 367)

While such three dimensional imaging practices were already in wide and diverse use, the consortium was concerned that, as it stood, the images of human bodies emerging from such practices only existed in anatomic fragments. CT scans could provide volumetric data regarding skeletal structures, body sections could be anatomised and turned into volumetric images, but each of these data sets derived from different bodies, were imaged using different modalities and registration protocols, and hence lacked commensurabilities. Cross-sections could vary in depth by a centimetre or more, and images varied in degrees of resolution. Translation from one image order to another was hence uncertain (Spitzer *et al*. 1996). This disorder was compounded by the absence of an agreed, singular, anatomic 'norm' in this new medium, some stable point against which degrees of pathology could be visually determined. The consortium therefore concluded that 'no spatial data set of anatomic coordinates for the complete human body exists in the public domain' and recommended that the NLM support

> The development of an image data set of an entire human male and female ... such a Visible Human Project would entail the acquisition of corresponding computed assisted tomography (CAT), magnetic resonance interferometry (MRI), and cryosection images of a representative, carefully selected and prepared male and female cadaver.
>
> (Ackerman 1992: 367)

Three years later, in 1991, a research team was contracted to find and process suitable cadavers. The team named itself the Center for Human Simulation – 'a new entity emerging from the synthesis of three-dimensional imaging and human anatomy' dedicated to the production of 'a three-dimensional, high resolution database of human male and female anatomy ... derived from direct analysis of anatomical specimens'. The goals of the Center include the development of 'interactions with computerised anatomy in virtual space' and the education of 'the anatomist of the twenty-first century, a new type of anatomist/computer imaging specialist'.[9] The Center had access to more than three thousand cadavers a year through the services of three State Anatomical Boards, but because the cadavers were to be used as exemplars of normal anatomy the process of selection from among all the available donations was slow. Over a period of nearly three years, several cadavers were screened for their suitability. Screening was effectively designed to exclude the possibility of any visible pathology, anything which might detract from the project's claim to present strictly healthy and normal anatomies. Cadavers were screened

> For evidence of infectious or metastatic disease, surgery, or any other condition that might have altered or distorted the cadaver anatomy, or otherwise rendered it unsuitable for the project. The specimen's physical state was carefully examined for evidence of scars or distortion. The size and weight of the candidate were noted, and obese or emaciated cadavers were rejected. ... After the history and physical condition of cadavers was evaluated, radiographs of the entire body were taken of those judged acceptable. If no abnormality was found in them, survey CT images were obtained of each region of the body. MRI images were also acquired, initially of the head and neck and later through the entire body. The total information thus collected for each sample set cadaver was transmitted to the Visible Human Selection Panel established at the NLM.
> (Spitzer *et al.* 1996: 119)

The first cadaver to pass this rigorous audition was that of Joseph Jernigan. Before his execution, at the prompting of a prison chaplain, Jernigan had signed a donor consent form, making his corpse available for scientific research or medical use. Most donors provide organs for medical transplant, a use ruled out in Jernigan's case by his mode of death, by the poisoning of his tissues. Instead his corpse was selected, despite an appendectomy and a missing tooth, to be the first Visible Human Male, the first man in digital space. Reversing the logic of H.G. Wells's *Invisible Man* (1897), the man who could render his body transparent and inaccessible to vision, every structure and organ in the interior of Jernigan's body was about to become an object of exhaustive and globally available visibility.

The transformation of body into anatomical text has always involved a considerable technical labour. In earlier times Jernigan's body would have been subjected to the processes of classical dissection: his corpse preserved in formaldehyde, refrigerated, and layers of skin, flesh, muscle and sinew peeled away to reveal the internal structures to the eye. To transform his body into digital data required a new set of techniques. Within hours of his execution, Jernigan's body was airfreighted to Colorado, and the process of transforming the cadaver into the first Visible Human body began. This process involved imaging the body through a number of different media. The body was placed in the Center's MRI machine and fully scanned. This first imaging process was to provide a template for the intact body, which was to be digitally 'duplicated'. It was then frozen in blue gelatine at $-85°$ C. Once suitably solid, it was cut into four sections and each section was CT and MRI scanned.

After that, the body was systematically and very finely sliced into oblivion, the slices falling at the same points at which the CT scan had optically 'dissected' the body. Beginning with the feet, the frozen sections were fitted into a dissection device, a cryomacrotome, described by its creators as a 'milling device', which planed the cadaver at 1 mm intervals. After each planing the cross-section of the remaining body section was digitally photographed, so that each photograph registered a small move through the body's mass. Each of these photographs was then converted into a computer data file, and their position in the overall body registered according to the initial template. This technique effectively obliterated the body's mass, each planed section dissolving into sawdust due to its extreme desiccation. The process took nine months.

In this way the corpse was converted into a visual archive, a digital copy in the form of a series of planar images. The entire Visible Male archive consists

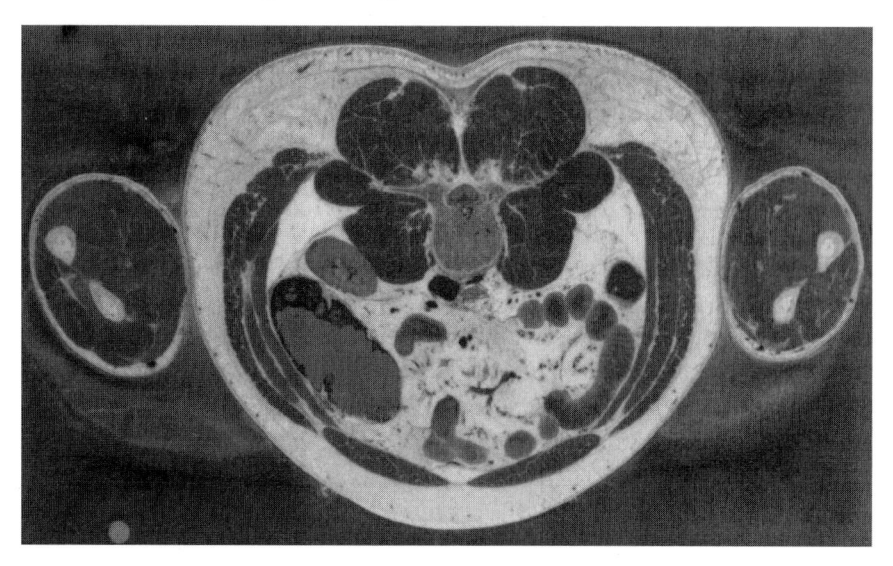

Figure 1.5 Section through Visible Male: thorax (image courtesy of National Library of Medicine, USA)

THE VISIBLE HUMAN PROJECT

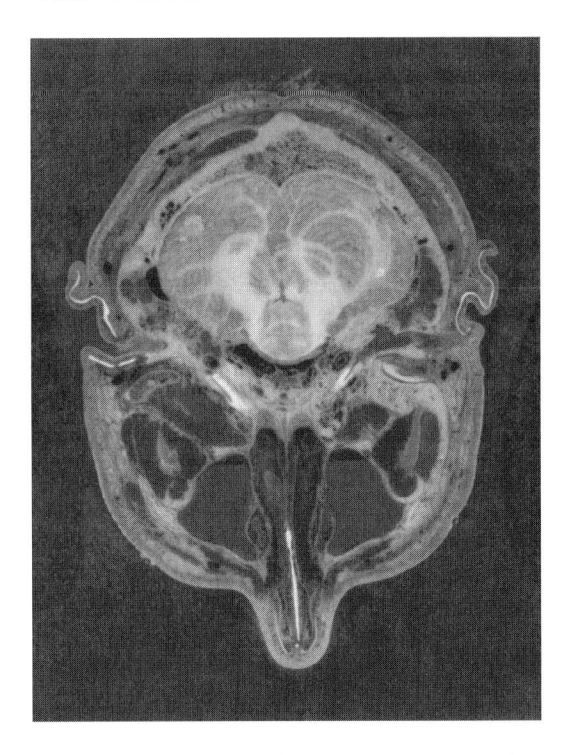

Figure 1.6   Section through Visible Male: skull (image courtesy of National Library of Medicine, USA)

of 1,878 24-bit digitised images of slices, and occupies 15 gigabytes of computer storage space, the equivalent of 23 CD-Roms. The Visible Woman represents a technical improvement on the Man: the body was planed into much finer cross-sections (5,189 sections) which produced higher resolution in the resulting images and a much larger data file. The files can be viewed one by one, showing highly resolved, transverse cross-sections through the body in realistic, photographic colour. (See Figures 1.5 and 1.6.)

Due to the animating and volume-rendering capacities of computer vision this archive is not merely an inert series of images. The slices can be restacked so that the appearance of volume and solidity, surface and depth, is restored to the virtual body. This restacking capacity enables unlimited manipulation of the virtual corpse. As one commentator describes it,

> The ... data set allows [the body] to be taken apart and put back together. Organs can be isolated, dissected, orbited; sheets of muscle and layers of fat and skin can lift away; and bone structures can offer landmarks for a new kind of leisurely touring.
>
> (Ellison 1995: 24)

THE VISIBLE HUMAN PROJECT

Blood vessels can be isolated and tracked through virtual space, users can 'travel' the entire length of the spinal cord or the oesophagus, and the body can be opened out in any direction, viewed from any angle and at any level of corporeal depth. Users can move around the body using hypermedia links and 'flythroughs', which allow navigation through the body's lumens from a vantage point similar to that of a tiny spacecraft. The body appears like a terrain which moves beneath and beyond the user's point of view. Blood vessels or intestines can be traversed as if the user is flying through narrow tunnels. The skeleton can be travelled as if flying at low level through a forest of bones. (See Figure 1.7.)

Moreover, the virtual corpse can also be animated and programmed for interactive simulations of trauma, of human movement, of fluid dynamics and of surgery. The heart can be made to beat, the veins to bleed, the flesh to bruise and lacerate. At present programmers are working to give muscles and bones different degrees of resistance and density in relation to haptic data transmitted to a surgical data glove, to be used by a surgeon performing simulated surgery.

The VHP figures are made available by the NLM as self-defined databases, indexed sources of complete anatomical information which can be downloaded from the internet to licensed users. The kind and number of current users are far greater than the NLM Board of Regents anticipated at the project's conception. Initially, the Board considered the primary uses of

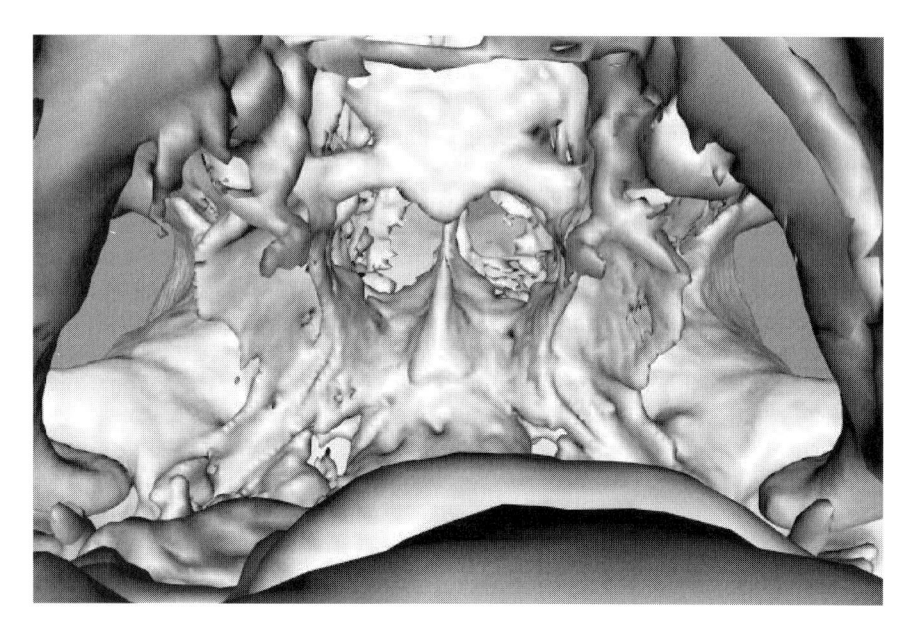

Figure 1.7 Flythrough of Visible Male bone surfaces (image courtesy of B. Lorensen, GE Imaging and Visualisation Laboratory)

the data would be for limited educational applications – the production of a small number of multimedia anatomy textbooks for medical students and high-school biology students.[10] At the time of writing, four years after the launch of the Visible Male data, more than 1,000 organisations and individuals in 41 countries hold licences to use the data set.

Many licensees are, as anticipated, multimedia companies which have produced CD-Rom versions of the cross-sectional VHP data, for use in schools and by a general audience. Many others are based in computer science departments in universities throughout the world, which have focused on the task of rendering and labelling the data, making it usable and intelligible for medical staff and students. The task of rendering the data, that is, generating volumetric anatomical images from the cross-sectional data, is highly specialised and labour-intensive, and many sites have focused on rendering one part of the body. The National University of Singapore site, the Centre for Information-enhanced Medicine, is focusing on the production of a brain atlas, while the University of Michigan's Scientific Visualisation Laboratory has produced browsers for the Visible Human Female head and pelvis, and the University of Pennsylvania Medical Center has produced an interactive knee programme that allows users to view and manipulate knee-joint data specifically. Other sites have focused on the production of flythrough software and animation, which allows directional navigation through lumens in the digital body, travelling along the colon, oesophagus or blood vessels.[11] Much of this rendering is directed towards the production of education software for use in teaching anatomy to medical students, but other kinds of VHP work is being carried out for clinical applications or research, in areas like biomechanical modelling – the simulation of musculoskeletal systems; virtual diagnostic protocols – the production of visualisation software for gathering and formulating patients' CT or MRI data; and surgical simulation and rehearsal (programs which allow surgeons to plan and virtually enact surgical procedures on the VHP data prior to carrying out surgery with a living patient).

Most of this work to date has been carried out on the Visible Male dataset. The Visible Woman dataset has been primarily used for reproductive anatomy, for example in the use of the images for virtual reality training in gynaecology, while the male data is considered able to adequately represent the 'human' body. As Cartwright (1998) reports, the Visible Woman has been generally less successful than the Visible Man, less of a media *cause célèbre* and less used as an anatomy text, despite the higher resolution of the data. This relative lack of interest in the VW data relates in part to her anonymity – her figure does not present the same narrative opportunities as Jernigan's. It also relates to her age and her post-reproductive status. As Cartwright describes,

> In at least one respect [the] particular configuration of the Visible Couple doesn't quite mesh with the larger family picture imagined by the [project]. The discrepancy hinges on the age-appropriateness of the Visible Woman for her role as partner to the Man. Whereas the Visible Man is based on the corpse of a 39 year-old, the Visible Woman's source was 59 when she died. ... Where the Visible Man's age is mentioned, it is cited as evidence of his status as exemplar of normal (healthy and fit) male anatomy. ... The Project's presentation of the Visible Woman suggests a different criteria for normal female anatomy. The Visible Woman is represented as older, her age is linked to her sex and reproductive function, and specifically it is implied that she is menopausal.
>
> (Cartwright 1998: 30)

This 'problem' with the VF data derives from one of the central epistemological problems in anatomy generally: the problem of the norm and the ordering of the distinction normal/pathological (Waldby 1996). In this case the age of the Visible Woman's body renders its status as normal problematic, departing as it does from an implicit equation of the normal female body with the youthful and the reproductive. Consequently the NLM plans to image at least one other, younger, female body, in order to provide a 'pre-menopausal' model for VHP users, as well as an infant or foetus, a move which would, as Cartwright puts it, 'make the Visible Family a viable reproductive unit' (Cartwright 1998: 30).

As I observed earlier, interest in the project and its spectacular visuality has not been confined to medical or scientific circles. Current and potential users include visual artists, the US military, car designers, movie producers, school children and animators, as well as more conventional medical personnel. Here I would not like to make too strong a distinction between what the scientific and nonscientific communities find so compelling about the data. While those who draw upon the project for research, clinical planning and management or anatomical pedagogy may regard the VHP as simply another rational medical tool, medicine itself is hopelessly enmeshed in a gothic cultural imaginary around life, death, reanimation, and the status of the corpse. Its tools cannot be separated out from this imaginary order, and, as I shall demonstrate in the course of this book, the VHP indicates the extent to which medical rationality is driven by confusions between animation and reanimation, life and the illusion of life, medical corpses and uncanny spectres. If medical interest in the project is motivated by pragmatic concerns and applications, it is equally motivated by the kinds of fantasmic mastery promised in the looking-glass world of the virtual screen, where the laws of everyday matter – the irreversibility of time, the inescapability of decomposition, the finality of death – are suspended.

ORGANISATION OF THE TEXT

The six chapters which follow pursue the Visible Human Project along several different lines. Chapter 2, Posthuman Spectacle, is a first attempt to open this image-object out to a theoretical traversal. The VHP is one of the most recent effects of what has been termed the 'cybernetic turn' in biomedicine and the biosciences more generally, the calibration of living entities according to the logics of the computer. In this chapter I begin by using Heidegger's questioning of technics to think through the particular kinds of bioproductivity that the cybernetic turn creates, both in relation to the VHP and more generally. Taking up his notion of en-framing, and the technical configuration of the world as standing-reserve, I read his famous essay 'The Question Concerning Technology' (1977) against the grain of its apparently anti-technological biases. I use some of his proposals to argue that the bioproductivity of the VHP and other biotechnologies rests not on a simple expropriation of a given nature but rather on the open-ended participation of the natural world in technology. Biotechnology could, in general, be considered as a mode of instrumental 'address', an exploratory form of intervention which is also an invitation, soliciting an active compliance from the productive capacities of living matter. Biotechnology seeks to instrumentalise the already instrumental capacities of living entities along particular lines.

In other words, biotechnology is a means of gearing the material order of living matter, and biomedicine in particular seeks to produce what I term 'biovalue', a surplus value of vitality and instrumental knowledge which can be placed at the disposal of the human subject. This surplus value is produced through setting up certain kinds of hierarchies in which marginal forms of vitality – the foetal, the cadaverous and extracted tissue, as well as the bodies and body parts of the *socially* marginal – are transformed into technologies to aid in the intensification of vitality for other living beings. The biovalue specific to the VHP derives from the instrumentalisation of corpses and their transformation into putatively normative human archives. As digital archives the Visible Human figures partake of the visual and distributive economies particular to digital data, which lend them their particular usefulness as norms for the burgeoning field of digitally based medical imaging. The VHP also stands in a strong family relationship to another computer-based archive, the Human Genome Project. In the second part of this chapter I discuss these two archives in relation to the idea of the posthuman. Both archives present the human as a delimitable and specifiable content, yet both demonstrate that any such presentation rests on a prior calibration of bodies according to the machine logics of the archive itself. Both archives stage human species-being through the most thoroughgoing instrumentalisation of the human as itself a system of technics – an array of visual or genetic data. The human as essential and discreet content only becomes possible through its utter technological contamination. Contrary to a whole line of argument that has become associated with the figuration of the cyborg, I argue that this techno-contamination is not strictly associated with the cybernetic turn but is rather conterminous with the human, its technogenic network of production. The posthuman, in my reading, is neither a figuration nor an evolutionary moment. Rather it is an effect of the slippage involved in effacing or naturalising this network of production in the interests of maintaining the *amour propre* of the human, moments of disjuncture which leave this technogenic network exposed and available for critical analysis. The Visible Human Project is only one of many such moments, which are thrown up by abrupt innovations, and the lag between an innovation and its domestication.[12]

Chapter 3, Theatres of Violence: the anatomical sacrifice and the anatomical trace, investigates just such a moment of slippage, and also provides a prehistory for the VHP. It turns to the early history of anatomy, a history which the VHP recapitulates within its own apparatus, and lays out some ways in which the entire anatomical project, so important to the figuration of 'Man', could be read as posthuman. It does this in two ways. First of all it discusses the hierarchies within the category 'human' that designate some bodies as useful matter, to be anatomised in the interests of others. Medicine has a long history in which certain persons are used as test technologies or instrumentalised to produce a surplus value of knowledge, in order to enhance the well-being and functioning of the more fully human. The choice of Jernigan as the material for the first Visible Human refers the VHP back to a whole history in which the 'book of Man', the anatomical archive, is produced through the sacrifice and dismemberment of those cast out of the social contract, criminal subjects as well as other racially marked or socially marginal bodies.

Second, this chapter examines in some detail the extent to which even the earliest anatomical specifications of 'Man' as perfectible species-body do so through a machine analytics and the productivity of certain kinds of writing technologies. The discipline of anatomy involves the specification of macro-anatomical morphology as a given and coherent organic assemblage. Nevertheless, that morphology can only be produced through the spatial logics and inscriptive imperatives specific to the anatomical Atlas. Until there is an Atlas this is no anatomical morphology. It can only be brought into existence as a morpho-spatial entity through the particular graphic and bibliometric trajectories of the Atlas, which specifies the multifarious involutions and folding self-contiguities of the corporeal interior as a homogeneous and quantifiable spatial order, a rational and mappable space. Consequently, the move from a book-based to a virtual anatomy signalled by the VHP is not the porting of a given organic content from one neutral, descriptive medium to another, but rather involves working living bodies (clinically, surgically) according to entirely new orders of morphology and spatialisation. This chapter compares the forms of corporeal topography produced through bibliographic Atlases with the tomographic morphology produced by the VHP, in order to demonstrate the centrality of the anatomical trace system and its particular mediatic qualities in the production of surgically operable living bodies.

Chapter 4, Virtual Surgery: morphing and morphology, is in many ways a continuation of Chapter 3, and attends to the multiple ways in which the VHP alters the ways that living bodies can be technically addressed. This chapter describes the rapidly expanding range of clinical, pedagogical, surgical and ergonomic applications for the VHP. The Visible Human data are currently being used as exemplary test data for the development and refinement of algorithmic methodologies for surgical simulation, virtual imaging methods like virtual endoscopy, and surgical planning and rehearsal using patient data. The data are being used as a set of *operable images*, able to be deployed as surrogates for actual organs within simulated surgical procedures. As such, the images lay out certain kinds of possibilities for surgical navigation and transformation of living bodies, and can be used in an open-ended way to speculate about the possible biomedical traversals and transformations of macro-anatomy. Above all they lend themselves to the desire that the gothic, fractal, multiple involuted and self-contiguous interior of the body can be exhaustively mapped into a rational, homogeneous and quantified space. This desire is examined in relation to the anatomical 'flythrough'.

This chapter also continues the themes of the previous chapter in that it is an extended consideration of the relationship between the spatial capacities of computation, and ways in which this produces certain forms of surgical and clinical orientation and navigation. The VHP is one of the primary visual objects associated with the growing preference for telematic surgery – the working of the bodily interior as remote space, as in keyhole surgery, rather than through the opening out of that space to the light of the surgery and the direct gaze of the surgeon. This chapter compares the kinds of surgical space produced by the conventional x-ray with the tomographic space produced by the VHP and other computer-based imaging modalities. Finally it considers the importance of the VHP as a normative text which facilitates the production of tomographic patient data along particular standardised lines.

Chapters 5 and 6 undertake a genealogy of the ways in which the VHP disturbs and redistributes the spaces of the living and the dead. Chapter 5, IatroGenesis: digital Eden and the reproduction of life, is an interrogation of the uses made of Genesis iconography within the project. The two figures are persistently referred to as 'Adam' and 'Eve', both by the project scientists themselves and by the popular media. I argue that this usage is not simply cliché, but rather articulates what I term 'IatroGenic desire' – a desire inscribed in biomedical and other biological technologies that seek to produce living entities as stable, self-identical commodities. This desire is discernable in other areas of biotechnics, notably in the more reductionist uses of genetics which attempt to establish a stable relation of cause and effect between genetic code and morphogenesis.

The deployment of Genesis iconography in the VHP indicates the extent to which the appeal of the Visible Human figures, both for medicine and for nonspecialist audiences, derives from an equation of digital code with vitality. The figures are widely treated as vital entities, dead bodies resurrected through the negentropic power of information. This attribution of vitality locates the Visible Human figures in what, following Cubitt (1996), I have termed the cybernatural Eden of computed space, with its technogenic ability to create an informational second nature. The chapter undertakes an extended investigation of the implication of life with information and organism with code. The practice of Artificial Life is considered as a privileged site for the production of cybernatural space. In A-life the computer is treated as a technology which can literally write life as code strings and set living entities in train within its purely negentropic, cybernetic environment.

The VHP cannot make such straightforward use of the equation of information with life, however, because it must negotiate a more complex and paradoxical creation story. I argue that the Genesis narrative is somewhat attenuated in the case of the VHP because the figures must retain a relationship of verisimilitude to the cadavers used in their production, in order to claim the truth status particular to anatomical texts. Hence as reanimations or resurrections, the VHP figures articulate certain tensions within the biomedical approach to death. One facet of IatroGenic desire, I argue, is a desire for bodies to behave as closed mechanical systems with reversible temporalities, rather than as non-reversible, chaotic systems which necessarily move towards death. The Visible Human data, as morphable images, enact such a reversible temporality within their aesthetic, bodies which can be dissected and resected without consequence. IatroGenic desire also works off the ambiguity of the term 'reproduction', with its connotations of both sexual reproduction and self-replication. Hence to pose the VHP figures as Adam and Eve is to appeal to a moment of organic origin for the figures, while also treating the computer as a reproductive technology, which can reiterate this moment of origin in a series of 'good' copies.

Chapter 6, Revenants: death and the digital uncanny, examines in more detail the relationship between the VHP, biomedicine and death introduced in the previous chapter. It argues that the deployment of Genesis iconography discussed in the earlier chapter is not only a way of articulating a desire for stable forms of 'reproduction, but is also a defensive move, a means of pre-empting critical attention to the project's reliance upon corpses as primary sites for a knowledge of living bodies. All anatomy relies upon the corpse to model life, and hence presents a serious difficulty to biomedicine's explicit commitment to the clear demarcation of the living from the dead. Medicine's explicit project is the increase of human vitality and the accumulation of life against death. This project shapes multiple aspects of its practice, some of which — disease prevention, life-support technology, the equation of death with pathology — are discussed in this chapter.

At the same time, medicine relies upon productive encounters with corpses, donor cadavers, foetal tissue, and other forms of marginal life and near-life. Such entities act as research matter and as source of organs and tissues, crucial elements in the biomedical enhancement of vitality for the more properly alive. This implication of the living and the dead becomes more profound in medicine's continuing reliance on a concept of the living body as mechanical system. The idea of mechanism, with its predictable relations of part to whole and cause to effect, finds its most apt application to the corpse — an anatomical body from which the complications and open-endedness of subjectivity and vitality have been subtracted.

In consequence, the medical idea of life is haunted by the corpse, and medicalised forms of death seem to present possibilities for prosthetic life-after-death. It is this possibility which locates the Visible Human Project within a genre of gothic medicine — an iconography for terminal states and animate corpses. It also locates virtual space as a haunted site, its negentropic qualities put to uncanny ends. As reanimated corpses the Visible Human figures display the energy and paradoxical embodiment of the ghost, digital revenants who return from the post-natural space of data to haunt the public spaces of the web and the laboratory. In this way they figure the redistribution of the qualities of life and death, and the space of the living and the dead, effected by the vitality of information.

Chapter 7 brings together a number of these themes to speculate about the kind of collision the Visible Human Project iconography stages with the user at the screen interface. It gives a more extended speculation to the idea of the Visible Human Project as a post-natural mirror-image of the human, one which betrays rather than conceals a human indebtedness to technogenesis. It considers the possible ways in which the VHP's existence reconditions the spatial relationship between the actual and virtual sides of the screen, and questions the location of subjectivity and embodiment.

2
POSTHUMAN SPECTACLE

THE VIRTUAL *GESTELL*

If both public and scientific interest in the VHP is driven by fascination with the preternatural possibilities of cyberspace, biomedical interest is also driven by a set of practical concerns. The VHP represents the most recent solution to an old medical problem, the problem of bodily opacity. The entire anatomical enterprise, and its long historical project to bring medicine under the sign of science, is organised around the visualisation of pathology in the corporeal interior. Anatomical medicine concerns itself with the identification of pathology as interior topography, which must be visually encountered in order to be fully known, a geography which demands optical traversal. As Foucault, writing of enlightenment clinical anatomy, puts it,

> The [medical] gaze plunges into the space that it has given itself the task of traversing. ... In anatomo-clinical experience, the medical eye must see the illness spread before it, horizontally and vertically in graded depth, as it penetrates into the body, as it advances into its bulk, as it circumvents or lifts its masses, as it descends into its depths. Disease is no longer a bundle of characters disseminated here and there over the surface of the body ... it is a set of forms and deformations, figures and accidents ... bound together in sequence according to a geography that can be followed step by step.
> (Foucault 1975: 136)

Pathology is not a generalised state of being, a disease which afflicts the whole system, but a local and readable lesion, a mappable topography whose logic is spatial and visual. Hence the opacity of the body's exterior surfaces and tissues, its deflection of light and its material self-enclosure present one of medicine's persistent epistemological problems. This problem has been addressed by a succession of visual technologies which try in various ways to anatomise the body – dissection within the anatomical theatre itself, the anatomical atlas, the multifarious endoscopes which traverse the body's lumens, the radiograph and the more recent kinds of computed vision already mentioned. Each of these technologies transluminate the body in some way, open it to the incursion and projection of light or some other radiant spectra, so that its tissues become readable and interpretable as projected images, traces on a page or screen. Each technical innovation seeks to correct the 'deficiencies' of the kind of vision it replaces – the CT scan, by making a depthless optical 'cut' through the body's tissues it seeks to correct the superimposition of depths found in the radiograph, for example.[1]

The Visible Human Project is such an attractive object for biomedicine at present because it embodies a number of such corrections simultaneously, while taking full advantage of the abilities for modelling, animation and segmentation available in the virtual medium.[2] It counts as the present high point in the general application of computed vision to the problem of bodily opacity and body modelling in medicine, an application which is opening out new kinds of medical knowledge and research at a startling rate. In addition to the forms of vision already listed, computer imaging is used in new forms of surgery, where computers set laser beams to finely calibrated co-ordinates; in the many permutations of telemedicine, where the clinical and even surgical encounter between doctor and patient takes place remotely, through workstations, networks and specialised imaging and haptic technologies; in the use of virtual modelling as a form of medical research, where the model is the experimental object; and in medical pedagogy, where medical students learn anatomy, physiology and radiology through interactive programs at the computer workstation, rather than by perusing a book-based atlas. These and other applications in the clinic, the surgery, the hospital and the laboratory indicate the extent to which the computer screen has become the dominant way first-world medicine frames its object, the human body.

This framing takes place also in a second sense, in the positing of the body's molecular organisation as cybernetic system. This proposition has been articulated in molecular biology, immunology, endocrinology and other medical disciplines which understand the body as a network of informational systems, working through code, signal, transcription, interference, noise, and the execution of programmes. Morphogenesis and organism coherence emerge from DNA data, encoded instructions for the ordering of proteins, and diseases like AIDS proceed through viral reprogramming of the body's immune system. This framing has furnished biomedicine with some extremely powerful and productive techniques for instrumentalising molecular matter. New biotechnologies for recombinant DNA engineering, the insertion of genes of one species into the cells of another species, are enabling the production of what are literally new forms of life. Since the mid-1980s these techniques have been used to create transgenic species like the OncoMouse, a laboratory mouse genetically engineered to develop breast cancer, the first multicellular organism to be patented as a human invention (Vasseleu 1996). Other transgenic techniques are designed to grow human-compatible vital organs in the bodies of pigs or baboons, ready for harvesting and surgical transplanting into human bodies. Treatment for certain diseases now proceeds through gene therapy, which rectifies genetic 'deficits' or abnormalities like cystic fibrosis through the insertion of corrective genetic material directly into the patient's cells (Wilkie 1996). The techniques for mapping and rewriting genetic code are being consolidated and systematised through the Human Genome Project, which will, some time in the near future, complete its task of describing genetic 'Man'.[3]

Medicine's multiple adaptations of computer vision and cybernetic models have evidently provided it with new and precise kinds of leverage in its therapeutic projects. As we will see in the course of this book, the Visible Human Project alone has already generated synergistic effects in various fields of clinical and prosthetic imaging, enabling new forms of medical research, new forms of surgery and surgical training, new diagnostic practices and medical imaging benchmarks, which will no doubt preserve and enhance life and well-being for many. Victor Spitzer, professor of biology and radiology at the University of Colorado, and leader of the team that located and processed the two VHP figures, lists some of its applications:

> The images are being used for the teaching of anatomy and other health subjects from high school through to medical school. They are also being used as the basis for models of radiation absorption and therapy, crash therapy, ergometric design, and surgical planning. A virtual colonoscopy has been demonstrated, and a virtual laparoscopy simulator is expected shortly. The data set is providing the visual basis for surgical simulation. It is being used as a starting point by medical illustrators and is serving as a common source of images for the development and testing of rendering algorithms. The value of this national resource in the public domain increases through its applications. Its utility will continue to grow as related databases are attached to it, and as more attributes are given to its image elements.
>
> (Spitzer *et al*. 1996: 130)

In creating the VHP the scientists and technicians involved have utilised dead bodies in the production of multiple use values for medicine, an economy which characterises medicine generally and is generated by its biotechnical expertise. Medical imaging tends to rely on a rhetoric of verisimilitude and transparent communication to articulate its trajectories, to present its innovations as gains in accuracy, and its framing practices as gains in transparency, the passive situating of a re-presented organic object within an optical field. Hence the virtues of the VHP for medical education and research are frequently described in the quantitative terms of a strictly mechanical economy; as a modality which conveys *more* visual information than previous methods, *higher* resolution, *greater* realism and *better* detail. One bio-engineer, for example, situates the VHP within a history of steadily improving anatomical realism propelled by improvements in the media of demonstration, an ever improving carrier for an essential anatomical content.

> Anatomical atlases have historically consisted of artistic illustrations based on observations from multiple cadaver dissections. More recently, photographic atlases of cadaver dissections have gained favour due to their improved anatomical accuracy and detail. ... Much as photographic atlases added a higher degree of accuracy and information, the Visible Human data sets have provided a new level of dimensional and relational information to be realised.
>
> (Kerr *et al*. 1996: 57)

Nevertheless, the economic qualities of the VHP and other practices of biomedical imaging, their ability to yield a profit of knowledge and a reconfiguration of surgical and clinical practice, are generated not by *framing* in this passive sense but by *en-framing*. Here I am evoking Heidegger's term '*ge-stell*' (en-frame) to account for the *productivity* of this kind of medical imaging, its agency in the world and its ability to generate value. As will become evident throughout this book, the VHP, and the entire family of biomedical and bio-scientific images, are not simply re-presentations of pre-existing objects but rather what I will call 'operative images', images or trace systems which are in-themselves technologies, able to effect transformations in the bodily tissues that they seem only to re-present in the space of the screen. They are image systems which do not re-present a referent but rather materially implicate the world of natural objects that they stand in for. Heidegger's account of techno-science is also useful for thinking about the confrontation which the VHP stages with certain humanist values, a point I will pursue later in this chapter. Here I want to draw upon Heidegger's famous questioning of technics, in order to begin drawing out the techno-cultural significance of the VHP.

THE TECHNOLOGICAL FRAME

En-framing describes a particular way of producing technical objects, of bringing objects forth into presence, which Heidegger (1977), in his sustained interrogation of the 'question concerning technology' attributes to the operation of modern techno-science. Heidegger dispenses with the conventional reading of technology as 'applied science', and instead gives primacy to the operation of technics in the producing of scientific knowledge. As Latour (1990) and Idhe (1983), among others, have demonstrated, scientific knowledge is grasped only through the material trajectories of technologies and their partial instrumentalisation of natural objects, the particular ways in which instrumentation makes nature *apparent* and accessible to material reordering through magnification, calibration, spectroscopy and the like. Hence scientific perception is only possible through, and dependent on, technologies. These are the means through which relationships can be set up between scientists and the natural world.

Nevertheless, Heidegger's posing of the question of technology is not addressed simply to technology as instrumentation, although this is one of the forms of technical praxis. The most banal way to approach the question of technology is, he states, simply to assume its instrumentality, its means-end logic, which seeks to procure rational ends through the rational use of tools. Hence the logic of instrumentality is that of cause and effect, action which produces results. This construal of causality is problematic, in that it is already implicated in a particular technical teleologic, an idea of causality locatable in a particular, modern, technological moment. Heidegger opens up the term 'causality' by setting up a contrast between a traditional technology of *poiēsis* and an extractive, modern 'high' technology which 'challenges forth' the world. He refers to the classical Greek understanding of '*causa*', as 'that to which something else is indebted' (Heidegger 1977: 7), a network of co-responsible, interdependent relations of productivity. This kind of causal relation is exemplified in his account of the production of a silver chalice:

> Silver is that out of which the silver chalice is made. As this matter it is co-responsible for the chalice. The chalice is indebted to, i.e. owes thanks to, the silver for that out of which it consists ... the sacrificial vessel is at the same time indebted to the aspect (*eidos*) of chaliceness. Both the silver into which the aspect is admitted as chalice and the aspect in which the silver appears are in their respective ways co-responsible for the sacrificial vessel. [The third cause] is that which in advance confines the chalice within the realm of consecration and bestowal, [the *telos*]. Finally there is ... the silversmith [who] considers carefully and gathers together the three aforementioned ways of being responsible and indebted.
>
> (Heidegger 1977: 8)

This reciprocity of responsibility and indebtedness is not a moral economy but a productive moment, a particular way of bringing something *into* appearance, to occasion its arrival, an accounting for the presence of technical objects as such. As Weber (1996) glosses '*causa*': '[it] designates a relationship of being-due-to. This in turn involves not merely a private or negative relation: to be "due to" is to appear, to be brought into play thanks to something else' (Weber 1996: 63).

It is this co-responsible bringing-forth of entities which Heidegger terms *poiēsis*, a making present of things which is not limited to technical production but extends to the auto-productivity of living beings and to artistic production. *Poiēsis* is moreover a process of becoming visible for Heidegger, an event within an order of visuality where things come to appearance, *are revealed as the process of arrival*. Hence *technē*, the mode of technology which belongs with *poiēsis*, 'is a mode of revelation. It reveals whatever does not bring itself forth and does not yet lie here before us' (Heidegger 1977: 13). There is a tension here between 'revelation' as that which unveils an already existing, yet concealed entity, and revelation as that which makes present by making apparent, which only comes into being in being apparent, an apparition. Weber (1996) in his reading of 'The Question' introduces a further inflection here, contending that *'Entbergung'*, the term usually translated as 'revelation', is better rendered as an 'unsecuring', a 'putting in train' or 'harbouring forth' of something, a suggestion which strengthens the claims of the second form of revelation; something which takes place through a shifting, or a transmission of place.

Modern technology, for Heidegger, betrays a quite different economy in its productivity. It shares the capacity for revelation with *technē*, but puts this capacity to different ends.

> The revealing that holds sway throughout modern technology does not unfold into a bringing-forth in the sense of *poiēsis*. The revealing that rules in modern technology is a challenging which puts to nature the unreasonable demand that it supply energy that can be extracted and stored as such.
> (Heidegger 1977: 14)

Modern technology in this account is characterised not by a circular economy of reparation, a symmetry of debt, responsibility and co-productivity, but rather by its production of surplus value through the instrumentalisation of nature, 'driving out maximum yield at minimum expense' (p. 15). The auto-poiēsis of the natural world, its self-generating and self-organising qualities, are rendered as use values by the operation of modern technology, which for Heidegger is primarily extractive and parasitic, a perversion of the symmetrical economy of 'causa' into a unilateral economy of cause and effect. As Weber (1996) puts it, instead of traditional technics which brings-forth, modern technics is a 'driving or goading-forth: ex-ploiting, ex-tracting, ex-pelling, in-citing … not just an exacting but also an extracting of that which henceforth counts only as raw material' (Weber 1996: 69). The operation of modern technology is to order the world as use value and immediate resource, to make it knowable and accessible, ready to hand, through such ordering. This then is the process of en-framing; nature is *gestellt*, that is, 'placed' 'in the sense of being entrapped, maneuvered into a place from which there is no escape' (Weber 1996: 68).

This way of placing nature, of securing it and making it appear, is enabled through the calculative logic of techno-scientific knowledge, which Heidegger equates with physics. Physics produces nature as a 'standing-reserve', a manifold of cause–effect relations, a set of physical laws which are amenable to experiment and extraction precisely because of their calculability. 'Everywhere everything is ordered to stand by, to be immediately to hand. ... Whatever is ordered in this way has its own standing ... the standing-reserve (*Bestand*)' (Heidegger 1977:17). To know natural objects within the terms of en-framing is to place them at the disposal of instrumental rationality, to know in a way that is open to, and takes place through, calculation and technical manipulation. Physics is an interrogative knowledge which demands that 'nature reports itself in some way or other that is identifiable through calculation and that it remains orderable as a system of information' (p. 23). Hence techno-scientific knowledge is inextricable from a 'material gearing of the world' according to the operational conditions of technologies themselves. In this account scientific knowledge is never simply a question of adequation; it is always posed in terms which are articulable with the material forms and limits of instrumentation, a knowing *of* which is always a knowing *for*, meeting the need for analyses of nature that open up auto-poiētic entities to calibration and engineering, and to a general economy of commensurability. So if organisms are posed as systems of information for example, this knowledge works for the array of biotechnologies (e.g. polymerase chain reaction, viral vectors, 'reading' enzymes, gene databases) which are designed as informatic tools, which encode, decode, record or reprogramme.

Das Gestell also has connotations of visuality and of the telematic operations of technology that are peculiarly suggestive when considering the ordering of medical imaging technologies. As I have already pointed out, if the work of technics is a revelation which is also a venturing forth or unsecuring, it suggests forms of visuality which are also practices of dis-placement, rather than a simple and stable unveiling of what is already in place. The making over of the world as standing-reserve implies the creation of multiple forms of technical proximity, ordering the world so that it is ready to hand, re-placed or replace-able so that it can be technically accessible. In 'The Age of the World Picture' Heidegger (1977) gives this proximation a specifically visual inflection. Here he characterises the working of technology as performing the work of *vorstellen*, the 'setting-in-place-before that is an objectifying, a bringing to stand as object'.[4] This bringing-to-stand of the natural world as object takes place through the primacy of the visual which characterises scientific knowledge, the dependence on visual ordering and remote sensing which distinguishes scientific knowledge from, say, philosophical knowledge. This visual ordering is not a practice of re-presentation, a copying of the world as picture, but rather a practice of perception, an interpretive relation which takes place through a privileging of visuality. This visuality is set up by the organisation of modern technology which addresses the world *as picture*, and addresses the subject of technology, that is 'Man' as observer, one whose understanding is played out in and as vision. Hence much technology acts panoptically, as a machine which gathers up and orders the world visually and makes it available to be viewed as system before the viewer.

> Where the world becomes picture, what is, in its entirety, is juxtaposed as that for which man is prepared and which, correspondingly, he therefore intends to bring before himself and ... set in place before himself. Hence world picture ... does not mean a picture of the world but the world conceived and grasped as picture.
>
> (Heidegger 1977: 129)

So technology is that which assembles disparate locations and orders them as visual proximities. A privileging of vision is further implied in the term *ge-stell*, to en-frame, with its connotations of the window and the work of art, the flattening out and collapsing back of the world on to a flat surface, a screen. Contemporary scientific knowledge is replete with a seemingly endless array of remote visual sensing apparatuses which bring the recessive, microscopic, remote or obscured life of the world to the perception of the scientist through screen technologies. The Visible Human Project is one such apparatus, a means of bringing the obscured bodily interior into visuality. Other apparatuses – to borrow and expand upon Haraway's (1991: 189) list – include sonography systems, magnetic resonance imaging, scanning electron microscopes, computer-aided tomography scanners, satellite surveillance systems, micro-photography, endoscopy, interplanetary exploratory satellites, and so on. These devices are not simply adjuncts to scientific knowledge but are often the condition upon which knowledge can exist at all. As Heelan writes of quantum physics, for example:

> Quantum mechanics is unintelligible without the introduction of the scientist as an observer embodied in the empirical process of instrumentation and measurement ... the measuring instrument constitutes a readable technology, and the subject so extended is the body appropriate for a quantum mechanical observation.
>
> (Heelan 1983: 208)

The production of readability involves not only the production of the world as *Bild*, as systemic picture, but also the overcoming of remoteness, the telematic bringing-to-stand before of the image. *Das Ge-stell* is, in Weber's (1996) translation of the term, interpreted not as *en-frame* but as *emplacement*, a term which intimates the ways in which technics both arrests space and upsets a stability of space or place. The classical, Aristotelian, definition of place as 'immanence, stability and containment' is forced open by the working of modern technics, 'a way of unsecuring ... displacing [so that the] principle of containment no longer serves as the self-evident prerequisite of order. Instead, place as container "breaks up" and produces a "different kind of topography", *das Ge-stell*' (Weber 1996: 70–1). Modern technology, particularly televisual technologies, involves an incursion into the self-identity of place, making any locus available for multiple forms of 'broadcasting'. So *das Ge-stell* involves a framing that dis-places and emplaces, that redistributes place through the power of technics to challenge forth, to project or telegraph the natural world as image objects which address themselves to immediate perception and are brought within reach on and as screens, readouts, pages, radiographs and the like.

Read as an historical account of techno-science Heidegger's questioning of technology is highly contestable. Stengers and Prigogine (1997) have, for example, recently argued that attributing to all science an irreducible agenda of the manipulation and extraction of nature ignores the many arenas of research which address themselves to philosophical questions within a scientific tradition, and begs the question of how such a calculating science could ever be open to change in its ideas. Furthermore, to set up an harmonious economy of premodern technics against an extractive, parasitic economy of modern technics is to romanticise practices like pre-industrial agriculture, which frequently exploited the fertility of landscapes and precipitated famines or exhausted the fertility of soils (De Landa 1997). Heidegger's antimodern romanticism is somewhat problematic here because it suggests that any attempt to deploy his questions and speculations might simply recapitulate an antimodern technology position, a recapitulation I would wish to avoid.

Weber's (1996) reading of 'The Question' is useful on this point because he demonstrates, among other things, the untenableness of an absolute ethical distinction between premodern and modern technics, and between technics and nature, on Heidegger's own account. *All* technics, Weber argues, produce unsettling effects, asymmetrical, non-reversible economies, because all technics involve a bringing-forth of 'nature', whose innermost principle 'is its impulse to open itself to the exterior, to alterity' (Weber 1996: 67), its openness to technical forms of intervention and completion. As *poiēsis*, the natural world is an open system in which distinctions between technics and nature are highly relative, and the productivity of nature works artifactually, constantly proliferating its forms along lines which could themselves be considered as an evolutionary technics. The idea of the natural secreted in Heidegger's philosophy falls into a broad agreement with Dagognet's (1988) and Rabinow's (1996) antinaturalist readings of nature as *bricoleur* and as temporally unstable, spilled from state to state in concert with technics.

> Dagognet argues that ... nature's malleability offers an invitation to the artifactual. Nature is a blind bricoleur, an elementary logic of combinations, yielding an infinity of potential differences. These differences are not prefigured by final causes, and there is no latent perfection seeking homeostasis. If the word 'nature' is to retain a meaning, it must signify an uninhibited polyphenomenality of display. Once understood in this way, the only natural thing for man to do would be to facilitate, encourage, accelerate its unfurling: thematic variation.
>
> (Rabinow 1996: 108)

While for Heidegger this bringing-forth in conditions of modernity is a kind of perversion, this conclusion does not seem to me to either structure or exhaust the suggestiveness of his opening out of the question of technics and their relations to bio-economies of productivity and knowledge. I don't propose an 'application' of his account, but rather want to use it selectively to help draw out the question of the VHP, both here and throughout the text.

VISIBLE HUMAN TECHNICS

The VHP is a biotechnology, based upon the instrumentalisation of two dead bodies, which provide its raw material. Its product is intended to eventually assist in the instrumentalisation of an open-ended number of living bodies for therapeutic ends – the improvement of health and medical education, the more accurate imaging and removal of tumours or organ malformations, the planning and rehearsal of new surgical procedures, including keyhole surgery and telesurgery, and so on. It is then a technology which, like all biotechnologies, produces certain kinds of surplus biovalue, defined in general terms as a yield of biomedical knowledge and technique regarding the management and intensification of the productive and reproductive capacities of matter.

This term clearly stands in close relation to Foucault's term 'bio-power', which designates the enrolment of the forces of human life and embodiment in the formation of modern social order. Bio-power describes 'technolog[ies] of power centred on life ... the harnessing, intensification and distribution of forces, the adjustment and economy of [bodily] energies' (Foucault 1980: 144–5) in the interests of certain processes of social formation and regulation. My term biovalue makes less of a distinction between human and other kinds of embodied agents, and is not so specifically addressed to political processes as such. Rather it specifies ways in which technics can intensify and multiply force and forms of vitality by ordering it as an economy, a calculable and hierarchical system of value. Biovalue is generated wherever the generative and transformative productivity of living entities can be instrumentalised along lines which make them useful for human projects – science, industry, medicine, agriculture or other arenas of technical culture. Biotechnologies, the body of technique specifically concerned with the generation of biovalue, work to calibrate and order this transformative power, 'the elementary logic of combinations', in order to bring life into the realm of explicit calculations. Living entities (including human entities) so treated are open to systems of calculation and the leveraging of surplus values of health, performance, reproduction and knowledge. They can also converge with other economic systems. Currently the most productive forms of biovalue emerge from the calibration of living entities as code, enrolling them within bio-informatic economies of value which converge with capital economies. In the USA and some sections of western Europe, bio-technical research into the potentials presented by genetic manipulation, by tissue engineering (where computer-aided design [CAD] techniques developed for manufacturing industries are applied to the growing of tissue for transplants, skin grafts and the like), by xeno-transplantation, and the growing list of new methods for refiguration of living organisms, are increasingly propelled by commercial considerations and the profitability of bio-technical patents (Rabinow 1996). Biovalue is increasingly assimilated to capital value, and configured according to the demands of commercial economies.

The Visible Human Project can lay claim to such bioproductivity because of the new way it 'calls the corpse to account'. The history of anatomy is the history of such a calling, part of a broader medical project to rationalise death and to utilise marginal forms of life in its generation of biovalue.[5] The corpse is anatomy's primary resource, its raw material for the production of visual image-objects that describe the interior topography of dissected bodies. The VHP's efficiency and success lies in its innovative figuration of the corpse, its generation of an extraordinarily complex image-object which provides an apparently exhaustive spatial and topographical account of a whole body's substance. This account provides a unique kind of archive, a series of data files which record and store visual information about the two bodies anatomised. The VHP figures' coherence as archives is ordered through their referral to singular, once living persons, social identities which have been made into quantitative indexes. Moreover these transformed identities are treated as paradigmatic entities, archives which can serve as general accounts of the male and female Human Body.

It is this organisation as data archive which makes the VHP such an efficient 'standing reserve' for multifarious biomedical projects. It can be deployed as a static repository of commensurable images, viewed sequentially, or, as I have already begun to describe and will discuss in much greater detail in Chapter 4, it can be deployed as volumetric model and interactive surrogate. In either case this archive provides not simply 'descriptions' of body-objects, but rather *operative* images. They are, in themselves, a set of mathematical-Euclidean knowledges of bodily ordering. They provide an account of the possible ways that living bodies can be instrumentalised as macro-anatomies, made accessible to therapeutic or surgical technics.[6] As I shall explore in detail in Chapter 3, anatomical image-objects are technologies that visually sediment accounts of bodily form. They give an account of technically ordered bodily space and provide a set of spatial and temporal orientations for 'navigation' within that space. As Hirschauer describes the use of anatomical texts in relation to surgical practice, 'anatomical pictures document products of dissecting labour, and thus also provide an idealised account of what has to be done. They also provide a schedule for what is to be done, and what the "natural object" should look like' (Hirschauer 1991: 310). To put it another way, anatomical images are operative images in that they always embody their own possible forms of implementation, the procedures for moving from image-object to body-object and back again.

The spatialisation of the corpse, the cryosectioning already described, places this account within a general field of computer-screen orientation, a 'tomographic' spatialisation which is general to computed medical vision and which is examined in detail in Chapter 4. Moreover the corpse is 'called to account' *as an archive of visual data*, so that the VHP has its being as a data object. The VHP partakes of the general productivity of data economies, an anatomical object which seems to re-present the qualitativity of a volumetric, dense body in its aesthetic, but whose substance is a form of mathematical code. Consequently the terms through which it is made to appear, its quantitative mode of being made present, open it out to all the complex forms of exchange, segmentation, iteration, scale manipulation and transmission that are available in digital economies, forms of circulation only possible when a common code and a standard value are in play. As code it is not merely transmittable but has its being in transmission – the quality of telematicity is always/already inscribed in it, so that each instantiation of the project at a workstation is only an instance of an object which exists in circulation and distribution. It is this which makes the VHP a kind of 'shareware', a networked entity which can be dispatched via the internet, and which can 'take shape', be made to appear at any properly equipped 'terminal'. Hence its presence as object is itself paradoxical, an apparent stability and volumetricity, a modelling of the living body conceived as a self-identical, self-contained and containing volume, a stable relation of surface and depth, inside and outside, which is itself always in a state of discreet decomposition, the native state of digital data as discreet binary code. It is unsurprising then that the actual production of the VHP figures *as figures*, as volumetric icons with an apparent surface and depth, involves a considerable labour of registration, modelling, rendering, segmentation, the assignation of colour and so on (Tiede *et al*. 1996). This labour works over the 'raw' data, much as the cryosectioning worked over the 'raw material' of the corpse, domesticating it to produce intelligible photorealistic figures. It is through this paradox between its digital substance and its volumetric aesthetic that the VHP figures are made 'ready-to-hand' at the computer workstation, placed before the gaze and hand of the user and able to be summoned up or stored as needed. Hence information gleaned from only two bodies becomes repeatable and serial use values, performing the work of *vorstellen* through the computer's capacities for visualisation, iteration, transmission and networking.

This telematicity is further implicated in what I described at the outset as the VHP's disruption of a quotidian sense of space. Part of its appeal to a global public imagination resides, I think, in the multiple and paradoxical spatial states in which the VHP figures seem to exist. The project presents the viewer with the spatial homologue of a body. The Cartesian definition of a body is *res extensia*, that thing which takes up a stable space, which is itself place. In this instance the viewer encounters a body icon which has no specific relation to place, which takes place in multiple sites of storage and transmission. Moreover, the VHP figures seem to transgress eschatological orders of space. The VHP figures have been consistently interpreted as existing in a state of posthumous relocation, both a transport 'beyond' to the space of afterlife, and a return from the dead. As 'raw material' the substance of the corpses used in the project was so completely worked over that, at the end, all that remained of the remains was the visual trace, a trace which has its existence in virtual space. Such a relocation inevitably evokes the general cultural problem of the distinctions between the space of the living and the dead, a problem I pursue in Chapter 6. Moreover, the possibility that a dead body could be apparently relocated and *reanimated* in virtual space suggests that biomedicine has gained access to the virtual as a counter-intuitive space of post-natural and posthumous vitality, a space where the non-reversibility of everyday time might be arrested and redirected, the dead brought 'back'. This suggestion is discussed in detail in Chapter 5.

Finally, it is the VHP's mode of 'revelation', its way of being made present as virtual object, its way of being 'brought forth' from the corpse, which donates an uncanny negative 'presence' to the image objects, their sense of latency. Like many bioscientific images, the VHP figures are produced through a sacrifice of their referent. As I shall discuss in detail in Chapter 3, the scientific world is a 'trace world', where work is performed and value produced in the first instance through the production and manipulation of material residues – electron micrographs, autoradiographs, icon-images, chromatography and the like, ways of writing natural objects into iterable media (Lynch and Woolgar 1990). This graphic drive frequently demands the partial destruction of that which it traces, the killing and dismembering of laboratory animals, for example, in order to render the results of certain tests. Similarly anatomical knowledge, the rendering of bodily interiors as operative images, can only take place when the anatomical body is sacrificed, killed and dismembered.

In the case of the VHP this economy of sacrifice is particularly evident, the production of an extraordinary trace through the execution (in Jernigan's case) and abolition of fleshly substance. Hence there is a sense in all bioscientific media that the trace carries the force of a donated and negative 'presence', not in the sense that it simply recalls a presence elsewhere, representing a momentarily absent object, but rather that, like the ghost, the condition of the trace's existence is the death of what it makes apparent. This sacrifice gives the trace its status as operative image, its ability to be used as actant in the scientific world. This implication of negative presence seems to me to be acute in the case of the VHP, particularly in the case of the Visible Human Male, because of *the form of the trace itself*. As archive the VHP is a repository of commensurable images, a set of visual information about what is claimed to be a paradigmatic human body. As volumetrical, photographic data however it is also a *portrait*, the icon of a particular person who once lived and acted in the world. In the VHM this effect is amplified by the working of the name, Joseph Jernigan, which both acts to secure the integrity of the archive, as repository of a single coherent body, and links this archive to a social identity and a known history. Hence the VHP embodies within itself the latent status of all subjects, all citizens, as objects within the optical field of biomedicine, potential 'standing-reserves' and sources of biovalue. It is this point which ushers in the question raised in this book's title, the question of the posthuman.

HUMAN ARCHIVES: THE VISIBLE HUMAN PROJECT AND THE HUMAN GENOME PROJECT

The Visible Human Project is, like the Human Genome Project, an attempt to map out the topography of the human species as a coherent biological unity in a stable relationship of difference from other species categories. The HGP (also funded through the US National Institutes of Health) is a vast research project which co-ordinates the efforts of several genetic research laboratories around the world to map and sequence 'the human genome'; that is, to specify the compliment of genetic material in the set of chromosomes of the human organism. In the first decade of the new millennium this project should yield up an extensive human database – a sequence map of three billion base pairs and somewhere between fifty and one hundred thousand genes (Rabinow 1996: 97) which lays out the location of the information encoded in the human genome.[7] This 'Book of Man' will, it is hoped by the project scientists, count as a paradigmatic genetic characterisation of the human species, its capacities and vulnerabilities:[8]

> Encoded in the DNA sequence are fundamental determinants of those mental capacities – learning, language, memory – essential to human culture. Encoded there as well are the mutations and variations that cause or increase susceptibility to many diseases responsible for much human suffering.
> (National Research Council Report 1988, cited in Keller 1992: 294)

While the HGP tries to map the microstructure of the human species, the genetic instructions that are generally understood to tutor the unfolding of the body's morphology, the VHP maps the morphology itself, the gross anatomy of the male and female human body rendered as a database. Hence in both these Human Projects, the limits of the human as species is set out as a large yet finite information database, a spatial, graphic ordering which acts as a digital archive, retrievable through computer networks and readable at workstations. In both cases the computer itself sets out the material conditions of possibility for these projects, and putatively representative examples of 'the human body' are translated into terms that can be recognised and worked through computation. As genetic or visual information these human projects provide a map of a normative human body which is orderable according to the order- and information-processing ability of the computer. In the case of the VHP the bodies are currently being made searchable through the specification of three-dimensional spatial coordinates for each organ. Users can access the VHP archive and instead of downloading the entire body data, they can instead specify an organ and download exactly the voxels (volumetric picture elements) that constitute the organ information in the database. The archive provides the 'raw' data for complex anatomical simulations which emerge from the ability of the computer to treat visual information as structures of mathematical data.[9] Similarly gene sequence data generated by the HGP and other genetic projects can be accessed through genetic databases like GenBank, which not only stores more than 200 million gene sequence base pairs for humans and animal species, but also allows users to search, match and cross-refer within this information architecture. GenBank is described as 'an intricately cross-linked array of databases where investigators can ... search for similarities among gene and protein sequences, trace their evolution, and jump from sequence data to the relevant literature' (Waldorp 1995a: 1356).

In these ways the VHP and the HGP open the human body out to the search capacities, hyperlinks and cross-referencing systems of the computer archive. The projects are, in Heidegger's terms, emblematic 'standing reserves', orderings of 'the human body' which decompose them according to the logics of different orders of instrumentalisation (digital optics in the case of the VHP, gene-mapping procedures in the case of the HGP) and which in turn set up ways to lend particular bodies (surgical patients, foetuses) to biomedical prostheticisation. The Human Genome Project will, it is hoped, provide a map which will enable the identification of genetically-linked diseases (cystic fibrosis, Huntington's chorea) and the eventual production of treatments at the level of genetic engineering (gene replacement or transfer) and protein and hormonal pharmacology to control gene expression in particular diseases.[10] The VHP lends itself to the design of macro-anatomical interventions, new ways to precisely specify and traverse macro-anatomies in Euclidean space. Hence it produces new kinds of medical imaging, surgery, artificial limb design and the other morphological and orthopaedic innovations discussed throughout this book.

These two Human Projects are, then, instances where biomedical science treats the human as species, incarnate life located in the natural world. The *project* of these Projects, along with virtually all other forms of biomedical research, is to address the category of the human, not in the first instance as social *subject*, but as driven organic matter, as organ-ism. That is, they address the human as it is continuous with other forms of animality, incarnate life open to the self-organising and self-propelling capacities of living matter, and specifiable through the same modes of scientific knowledge which can be used to characterise the genome of yeast or mice for example.

In particular these Projects, like all biomedical projects, address the human as *morbid and mortal matter*. Medicine has an absolute concern to maintain the integrity of the human as distinct and superior species-being, yet this concern is premised on its detailed recognition of the distinction's absolute failure. At the cellular and molecular level the organic integrity of the human is already dissipated, a provisional concatenation of cell lines permanently open to other kinds of bacterial and viral life, and to posthumous disaggregation. Biomedicine concerns itself with precisely this openness, human bodies as entropic and wayward matter. It tracks the ways in which human flesh lends itself to the agendas of disease, becomes animated by carcinogenic vigour or used for viral and bacterial replication, made an agent in contagion, or is simply subject to formal dissipation, the tendency of matter to disorganisation through time. To draw on Dagognet's (1988) phrase, it addresses the body in its capacity as 'blind *bricoleur*', as a logic of possible combinations which lead off in any number of directions as mutation or disease. Such openness leads the body eventually to its own disappearance, its absorption into vegetal matter through and after death. In other words medicine addresses itself to the *failure of distinctions between human and nonhuman life*, the endless ways in which the 'human' in the term 'human body' is redundant, a failure of conceptual prophylaxis against this body's immersion in and indebtedness to the non-teleological, indifferent flux of the natural world. Medicine's humanist vigilance lies in its assiduous establishment and policing of the borders between the human and inhuman world of microbiology, and between life and death.[11]

At the same time these projects address themselves to another kind of incarnate openness – the body's openness to instrumentalisation. To address the human body as 'organism', as biotechnologies do, is to address it as, literally, 'a complex of tools'.[12] Biotechnologies address and incite organisms as themselves technics. In the case of the VHP and the HGP for example, both projects proceed by not only creating an archive of knowledge *about* the human body, but also by asserting that *the body is itself an archive*, an organic form of storage and replication. This is not simply a metaphor. Rather the procedures of both the HGP and the VHP literally make the body as archive. Both projects are ways to itinerise and index the human body as a finite content, open to multiple forms of ordering and modes of retrieval. The crucial point here is that, as Heidegger proposes, bio-scientific ways of modelling and understanding the order of living matter carry the means of instrumentalisation within themselves. While understanding the human body as database or information archive may be metaphorical at one level, at another this mode of understanding produces material practices that work the body as database. The instrumentation specific to the VHP and the HGP address the body so that it acts as an archive, laying it open as a source of bioinformation. The VHP cuts up and scans the body so that it can act as a visual and morphological archive, while the HGP has a large repertoire of molecular practices (polymerase chain reaction, automated DNA fluorescence sequencing, pulse field gel electrophoresis) that make the body's sequences of DNA intelligible and mappable. In other words, both the VHP and the HGP make bodies able to act as specific forms of archive. They provide the material conditions that allow bodies to work in this way, creating new modes of intelligibility for their interior structure and microstructures, and new ways of articulating organs as tools. The biotechnologies used in medicine (surgical instruments, pharmaceuticals, implants, diagnostic imaging methods) work by temporarily or permanently suspending any distinction between endosomatic organs and the exosomatic organs that comprise technology in general. Here I am following Innis (1984) in his characterisation of all technologies as exosoma – structures which 'substitute for, extend and compensate for' the endosoma, and which 'have their own trajectories – dynamic logics or vectoral paths [that predefine] the ultimate grounds for the historical variability of consciousness and the forms of perception' (Innis 1984: 68). If *all* technology implicates and supplements bodily organs (Innis gives the examples of 'microscopes, glasses, telephones, computers, aeroplanes, weaving machines, printing presses', and eventually language itself) then biomedical technologies give an extra depth to this implication, and in the process throw into question the viability of a distinction between a natural inside and technical outside of the organ-ism, or of setting out any defining limit or distinct interface between organs and technics. Biomedical technology takes place through absorption within the very cells and molecules of human bodies (HIV prophylaxis that prevents infected cell replication), through inducing alterations in hormones or enzymes (the contraceptive pill), in body temperature (Aspirin), or through prostheses which double or replace or create endosomatic organs (pacemakers, hip replacements). Its purposes are therapeutic, seeking to enter into and transfigure the organism's capacities for morbidity and mortality, to bring its waywardness back within the management of human culture and the provenance of the social subject through the engendering of health and the holding off of mortality.

The project of these Human Projects, in the sense of their *explicit intention*, is, in fine, to advance the specification of the human as species-being, in order to render the body as a resource for consciousness. Like all medical projects they work to retain the living body's agency on the cultural side of the putative nature/culture divide, at the service of the subject of knowledge rather than at the service of microbial life, entropy and mutation. As Foucault (1975) observes in *The Birth of the Clinic*, the anatomo-clinical method, the historical development of clinical medicine out of the dissection and anatomisation of corpses throughout the seventeenth and eighteenth centuries in Europe, is the first science of 'Man'. For the first time the human is conceptualised as object of positive knowledge, arrayed with other living objects in the realm of nature. Foucault's subsequent historical work documents the generalisation of the sciences of 'Man' throughout modernity, the generation of criminological, psychological and sexological discourses which position 'Man' as object. Yet this initial historical gesture, the first moment of becoming scientific object is, he suggests, propelled by the prospect that such knowledge would control death through encountering death, and the material organisation of the human body which led it towards death. He writes,

> It will no doubt remain a decisive fact about our culture that its first scientific discourse [i.e. anatomy] concerning the individual had to pass through this stage of death. Western man could constitute himself in his own eyes as an object of science, he grasped himself within his language, and gave himself ... a discursive existence, only in the opening created by his own elimination. ... It is understandable, then, that medicine should have had such importance in the constitution of the sciences of man – an importance that is not only methodological, but ontological, in that it concerns man's being as object of positive knowledge, [the] possibility for the individual of being both subject and object of his own knowledge ... medicine offers modern man the obstinate, yet reassuring face of his own finitude; in it, death is endlessly repeated, but it is also exorcised; and although it ceaselessly reminds man of the limit that he bears within him, it also speaks to him of that technical world which is the armed, positive, full form of his finitude.
>
> (Foucault 1975: 197–8)

Man as transcendent subject and point of origin comes into being doubled and split, as simultaneously subject and object of scientific, positive knowledge, a split which is set up and stabilised in bio-technical terms. So on this account, and on Heidegger's account of technics, Man as voluntarist subject is produced through the same movement that establishes the human as species-object, knowable as anatomy, a techno-organism potentially at the disposal of the mind. The struggle of biomedicine is to analyse and manage the species body so that it can work as a sound foundation upon which a voluntarist subject can erect itself. The armamentarium of medicine is designed to secure self-identity against engulfment by the crises of disease and the destabilising evidence of embodiment, pain and mortality. It guards against the body's potential for 'corporeal irruption into consciousness' (Shildrick 1996: 3) and extends the temporality of willed life against the non-time of death.

Rendering the body as scientific object for an interpreting subject is, then, to en-frame it, in Weber's (1996) sense of 'emplace', to put it in a secure place in relation to, and at the disposal of, that subject. Nevertheless this attempt to secure and place can never be finalised; the dynamism of modern technics itself, and its ontological priority to 'Man' as subject ensures that the operation of technics is always destabilising, that it continually 'upsets the setup' (Weber 1996). The subject is not the creator and master of technics but that which is constituted in and by particular technical configurations with the world, *placed* in relation rather than *placing* in relation. The subject is summoned up by the configuration of modern technics, a responding to a call, an interpellation. As Heidegger puts it,

> Enframing is that gathering together ... which sets upon man and puts him in a position to reveal the real, in the mode of ordering, as standing-reserve. As the one who is challenged forth in this way, man stands within the essential realm of Enframing. He can never take up a relationship to it only subsequently.
>
> (Heidegger 1977: 24)

The suggestion here is that technics exercises a determining logic, evident in the setting up of subject/object relations as such, and that modern technics is hence never simply at the disposal of 'Man' but sets Man up and replaces and displaces him in dynamic ways. Any possibility of transcendental subjectivity is questioned in such a movement, so that 'the boundary between the human subject and its networks of production' (Wolfe 1995: 35) is exposed and eroded. Hence subjects of knowledge are themselves always within the en-frame of the technical, and the security of the voluntarist subject is always imperiled by the driven nature of technology, the unfolding of its own imperative logics and the non-predictability of its effects and incitements, its ways of bringing forth the world. As Weber puts it,

> [Technics sets] in place, but the fixity of such place setting turns into a placing of orders that can never stop. The more technics seeks to place the subject into safety, the less safe its places become. The more it seeks to place its orders, the less orderly are its emplacements. The more representational thinking and acting strive to present their subject matter, the less the subject matters, the more it idealises itself as pure will, as the Will to Will.
>
> (Weber 1996: 74)

This slippage, the insecurity of placement, is at its most fragile, it seems to me, in the case of biotechnology understood as a mediation between two registers, the human subject and the human species. Bio-technics can be utilised in the institution of this internal split, used to set up a hierarchy between them where the species is placed at the disposal of the subject as resource. Yet it also continually undermines the stability of this hierarchy by changing its terms and opening it out to unforeseen risks. It is in this slippage and spillage that the intimation of the posthuman takes place.

POSTHUMAN MEDICINE

The term 'posthuman' has come to designate a loosely related set of recent attempts to reconceptualise the relationship between the rapidly transforming field of technology and the conditions of human embodiment. These attempts are, generally speaking, a response to the cybernetic turn and the vitalisation of information, provoked by what Pearson describes as 'the scale of disorientation and displacement created by the impact of computerisation, the rise of new forms of engineering and new modes of knowledge, the creation of artificial life etc.' (Pearson 1997: 4). The most literal adjudication of the term posits the demise of the human in the face of a technological evolution, its absorption into the new informational economies as obsolete organic matter (Terranova 1996).[13] Some other less apocalyptic and more nuanced critiques, primarily those associated with cyborg politics, also take the posthuman to designate an historical effacement of some precybernetic, 'organic' human figure. The problems with this temporalisation of 'the posthuman' will be discussed below. For now I want to use the term to evoke a general critical space in which the techno-cultural forces which both produce and undermine the stability of the categories 'human' and 'nonhuman', can be investigated. How are the self-identity and transcendent status of the human secured as the 'not-animal', the 'not-machine' and the 'not-embodied', and in what ways does the purity of these categories unravel and contaminate each other? How do changing modes of instrumentation, 'the translation of the world into a problem of coding' (Haraway 1991: 164) alter the material trajectories of human embodiment and open up changes in relations of body with world?

The field of biomedical technologies seems to be a particularly suggestive site for opening up such questions because of the ways in which, in general, they implicate all the forms of being against which a pure human identity might secure itself, and force an encounter between them. What I mean by this is that, on my previous argument about the Human Projects and biotechnics more generally, such technics address the human precisely as an incarnate openness. While on the one hand the Human Projects might present the human body as a finite and stable content, their practices assume and promote its openness to profound, technically driven transformation, and its indifference to differences between human and nonhuman.

In this regard the Human Projects, and effectively all other biotechnical practices, play out a paradoxical dynamic which Latour (1993) attributes to scientific modernity as a whole. The productivity of science, its proliferation of technical and bio-technical innovations, derives, he argues, from its willingness to engage in the hybridisation of what is generally considered to be different ontological orders of being. The work of science, on Latour's reading, is precisely the generation of networks of production which bind together the agency and material orderings of things, animals and humans, translating each into mutually reciprocal terms (code, electricity, biochemistry) which enable new mixtures to be generated. This work of translation describes all those technological practices which amplify the contacts between humans and the nonhuman realm of organisms and technics, and which assume and promote an interchangeability between human, animal and technological organs.

At the same time, the *amour propre* of humanism, convinced of human exceptionalism, refuses to acknowledge these movements of translation. Hence science also engages in the prophylactic work which Latour terms 'purification', mapping out domains of putative ontological and organic integrity (species, genomes, anatomies) in order to conform to the demands of humanist ethics. Latour writes,

> The word 'modern' designates two entirely different practices which must remain distinct if they are to remain effective, but have recently begun to be confused. The first set of practices, by 'translation' creates mixtures between entirely new types of beings, hybrids of nature and culture. The second, by 'purification', creates two entirely distinct ontological zones: that of human beings on the one hand; that of nonhumans on the other.
> (Latour 1993: 10–11)

Hence the Human Projects act through a rhetoric of human distinctness and exceptionalism to precisely undermine any possibility of such purity, a move which, as Latour puts it, enables the work of hybridisation to go on unrestrained by humanist ethical concerns. The HGP claims, on the one hand to sequence a discreet entity, the Human Genome, yet on the other it utilises genes sequenced from other creatures to fill in much of its mapping work. Rabinow (1996) reports that several organisms (particularly mice) have been designated as model systems for the human genome, precisely because of the indifference of much genetic code to species distinction. For genetics humans and animals are simply macro-anatomies through which slightly different genetic information flows.

> Many genes work in the same way, regardless of which living being they are found in. Thus in principle, wherever we find a specific protein we can know what DNA sequence produced it. This 'genetic code' has not changed during evolution and therefore many genes of simpler organisms are basically the same as human genes.
> (Rabinow 1996: 98)

Moreover this specification of the Human Genome has already opened up possibilities for transgenic work, where genetic material is further exchanged between humans and animals. In this regard the HGP constitutes a tacit acknowledgement that all organisms are open systems which engage in transversal and intraspecies genetic exchanges as well as filial, species-specific genetic lineages.[14]

The Visible Human Project for its part claims to present a coherent and exhaustive human anatomy, the sum of organic parts, yet it disaggregates that anatomy not according to any logic of organic integrity but rather according to the logics of tomographic visuality and display. The VHP is a technology for calibrating living bodies according to the capacities of computer-generated space, and facilitating their surgical or orthopaedic reworking (keyhole surgery, facial reconstruction, tissue engineering) through a linkage with data homologs. Living flesh and digital data are brought into workable and assignable relations to one another, so that data space and anatomical space are placed in a kind of enabling confusion.

Latour contends that the workability of the distinction between translation, the work of producing hybrids, and purification, the work of declaring and maintaining ontologically distinct boundaries between humans and nonhumans, is currently undergoing a crisis. This crisis has been precipitated by the sheer volume of hybridisation taking place in contemporary technoscience, and the publicity that this process generates. The recent proliferation of monstrous entities – 'frozen embryos, expert systems, digital machines, sensor-equipped robots, hybrid corn, data banks, psychotropic drugs, … gene synthesisers' (Latour 1993: 49) – have weakened the possibilities for denial and the assertion of distinct ontological boundaries. The system of purification is failing, and the proliferation of hybrid entities is increasingly demanding public conceptualisation beyond the terms of humanist bioethics.

This failure, the 'disorientation' described by Pearson (1997), has energised recent attempts within the humanities to theorise current modes of hybridisation inaugurated through the cybernetic turn in the logic of technics. Haraway's (1991) account of this cybernetic turn describes it as the facilitation of informational reciprocities between different organisms, and between organisms and technics.

> Communications sciences and modern biologies are constructed by a common move – the translation of the world into a problem of coding, a search for a common language in which all resistance to instrumental control disappears and all heterogeneity can be submitted to disassembly, reassembly, investment and exchange. … The world is subdivided by boundaries differentially permeable to information. Information is just that kind of quantifiable element (unit, basis of unity) which allows universal translation, and so unhindered instrumental power. … The organism has been translated into problems of genetic coding and readout … [understood as] biotic components, i.e. special kinds of information-processing devices.
>
> (Haraway 1991: 164)

This cybernetic turn, the restless configuration of the world as interlocking information systems, is emblematic of what Weber (1996) terms the ongoing character of technics, the driven nature of technology. As Haraway's description implies, the expanding field of bio-informatics, in which the

VHP and the HGP are key components, is propelled by an internal logic and dynamics. Like any technological system, bioinformatics is driven to incorporate and systematise whatever heterogenous phenomena are at the borders of its current domain, to eliminate sources of noise and resistance to its informatic logic.[15] This capacity for self-propagation and expansion has a force which carries the effects and implications of information technologies well beyond any particular context of invention or intention. It is this driven nature of technics which works to open up unforeseen possibilities and instabilities for entities within its field of possibility. In this sense technics do not work at the service of some pre-existing hierarchy of human subject/human species but rather work over any such hierarchy, modifying and upsetting existing sets of historical terms in its articulation.

These instabilities have been thematised most consistently through the figure of the cyborg in recent techno-cultural literature, the melding together of cybernetic system and organism which is celebrated in Haraway's (1991) 'Cyborg Manifesto', and taken up in any number of other articles.[16] The figure of the cyborg emerges from the proposition that, if the human can be refigured as informational system, then the boundaries between the human and its stabilising historical other, the machinic, become purely conventional and are open to infinite transgression. Haraway writes, in her now much-quoted essay,

> A cyborg is a cybernetic organism, a hybrid of machine and organism, a creature of social reality as well as a creature of fiction. ... Modern medicine is full of cyborgs, of couplings between organism and machine, each conceived as coding devices. ... Cyborg replication is uncoupled from organic reproduction. ... By the late twentieth century ... we are all chimeras, theorised and fabricated hybrids of machine and organism; in short, we are cyborgs.
> (Haraway 1991:149–50)

The figuration of the cyborg has presented certain opportunities to think the kinds of transformations associated with the cybernetic turn. At a certain level it has worked to make explicit the human debt to and inextricability from technics, its openness in principle to technical logics of replication and configuration. Nevertheless I would argue that to conceptualise the inter-ramification of machinic information and living entity as hybridised *figure* carries with it certain historical and ontological commitments which close down more questions than they generate.

It seems to me that to figure the cyborg as hybrid, a figuration already conveyed in the neologism cyb-org, is to still imply a stable, linear descent, a move from the less to the more complex, from the more to the less organic, from the human to the posthuman. The cyborg *figure* implies an identity, albeit a hybrid identity, a point of stability around which a new cyber-anthropocentrism can coalesce. The cyborg figure emerges from its literature with an entire genealogy, a history of origins, which neglects to fully problematise the status of the human prior to the cybernetic turn. The human is assumed, for the most part, to bring an organic contribution to the new cyborgian being, an organic content untroubled by the play of technics heretofore. So for example in *The Cyborg Handbook* (Gray 1995b) several essays provide an account of cyborg genesis which simply assume an additive logic, the adding of informational technics to a human body which is already in place, and which also acts as the conscious centre of its own self-design.[17] In many of these accounts the potentially destabilising action of the cyborg has been recuperated as another mode of technical self-augmentation, here located as a development emerging from the NASA space program and the US military-industrial complex.

Hayles' (1993) more nuanced account of the cyborg posthuman nevertheless also invokes serial, symmetrical figurations, and an attendant drive to locate a moment of posthuman origin. She writes:

> I understand 'human' and 'posthuman' to be historically specific constructions that emerge from different configurations of embodiment, technology, and culture. A convenient point of reference for the human is the picture constructed by nineteenth-century American and British anthropologists of 'man' as a tool-user. Using tools may shape the body, but the tool nevertheless is envisioned as an object, apart from the body, that can be picked up and put down at will … the kind of tool [envisioned by anthropologists] was mechanical rather than informational; it goes with the hand, not on the hand.
>
> (Hayles 1993: 79)

Hayles' point is that the humanist concept of 'Man' is one in which 'Man' and technology remain conceptually and ontologically separable, and where no loss is incurred or change is wrought through dependence on technologies. Within this framework technologies are understood to be positive, cumulative additions to the body and mind, which always result in an increased yield of mastery over the entropy and unpredictability of the body. Hayles locates the beginning of the end for this masterful subject of technology at the Macy Conferences, a series of colloquia held between 1946 and 1953 in the USA which involved an interdisciplinary dialogue between communications system designers, geneticists, electrical engineers, and information theorists, a dialogue precipitated by the new science of cybernetics. The participants of the Macy Conference,

> wavered between a vision of man as a homoeostatic, self-regulating mechanism whose boundaries were clearly delineated from the environment, and a more threatening, reflexive vision of a man spliced into an informational circuit that could change him in unpredictable ways. By the 1960s, the

> consensus within cybernetics had shifted dramatically toward the reflexivity. By the 1980s, the inertial pull of homeostasis as a constitutive concept had largely given way to theories of self-organisation that implied radical changes were possible within certain kinds of complex systems. Through these discussions the 'posthuman' future of 'humanity' began increasingly to be evoked ... the posthuman implies a coupling so intense and multi-faceted that it is no longer possible to distinguish meaningfully between the biological organism and the informational circuits in which it is enmeshed.
>
> (Hayles 1993: 80)

Hayles, and other historians of information, are without doubt exploring a pivotal moment in the history of technics, a moment which can, I think, be productively interrogated through an articulation of the possibilities of the posthuman. Nevertheless, casting the posthuman as a figure, an ontology which replaces one category, 'the human', with another, the 'becoming cyborg', in a stable, linear movement, is only possible if Hayles' account of 'Man' the tool maker and user is taken as an exhaustively *descriptive* account of the earlier conditions of subjectivity and its relationship to technics. It involves accepting the proposition that, prior to the cybernetic turn, the relationship between 'Man' and technics was one of ontological separability, and that the human was never included in a nature which, I have argued following Heidegger, is already a practice of technics and an openness to technics. As Kirby (1997) points out, this posing of an originary moment of human purity before the fall into technics is a rhetorical move which flows from the logic of hybridity in general.

> It is against the unity of 'the before,' the purity of identity prior to its corruption, that the cyborg's unique and complex hybridity is defined. ... Haraway's 'disassembled and reassembled' recipe for cyborg graftings is utterly dependent upon the calculus of one plus one, the logic wherein pre-existent identities are *then* conjoined and melded. The cyborg's chimerical complications are therefore never so promiscuous that its parts cannot be separated, even if only retrospectively.
>
> (Kirby 1997: 147)

The cyborg figuration of the posthuman involves an acceptance of an organic origin for the human which is subsequently contaminated by technicity. As a proposition it is still committed to a norm of organic integrity and organic origin, which necessarily recapitulates an anthropocentric ideal of the human as essentially separable from and ontologically prior to technics, the human as a point of origin.

Rather than attempting to figure the posthuman otherwise, to confer it with some other origin, content, identity and morphology, I want to propose it instead as a particular kind of critical moment. The posthuman can best be understood as a point of view or insight made available by the contingency of technics, its driven and ongoing nature, and the incalculability of its consequences. The possibility of the posthuman is not to do with the transcendence of the human, its replacement, but rather with the recognition and exposure of the networks of production which constitute human techno-genesis.

The possibility of posthuman criticism has become available now, it seems to me, as a result of the extent to which the cybernetic turn in the logics of technics has driven earlier naturalising accounts of the human before it, into a state of disarray. This disarray is however nothing new. I would suggest that any transformation in technologic, particularly biotechnologic, will risk the putative stability of the human and send it out of phase, into non-coincidence with the existing terms of its naturalisation. New bodies of technique produce new forms of commensurability, new calibrating networks, which in turn lend themselves to particular ways of materially gearing the world. Shifts in forms of calibration change the trajectories of technics in which the human takes place, the operational conditions of its production and reproduction. They open it up to unforeseen possibilities for new modes of embodiment and translation, extension, supplementation and loss. Transformations in repertoires of technique introduce discontinuities and open-ended forms of instability into the material conditions which locate the human, shifting the terms of its enablement and disablement. The human is hence a category and status in a constant play of discontinuous mutation and provisional restabilisation, which must work with whatever field of technics is present.

To consider medicine from this point of view is to make evident its centrality in both precipitating such destabilisations and effecting forms of renaturalisation for an essential human status. As I have already demonstrated, medicine engages in a constant, paradoxical quest to shore up the status of the human through its material reconfiguration. Biomedical research seeks to work the body in more and more detail, at smaller and smaller levels of scale and at more profound depths of intervention. It configures endosoma as exosoma, making every partitioning and involution of the corpus available to its extended repertoire, demonstrating the relativity of surface to depth. Medicine has throughout its history found therapeutic, diagnostic or topographical applications for every new body of technical innovation – electricity, photography, optics, pharmacology – finding methods of bringing human bodies into line with their material trajectories. In the following chapter I examine medicine's early application of the printed atlas in the task of anatomical topography, an application which involved bringing the volume of the corpse into relation with the volume of the text. In this way, through its relentless application of new technologies to the body, medicine is one of the primary engines driving the expansion of particular technical logics. It has been quick to extend and exploit the new forms of imaging and corporeal engineering made available by the cybernetic turn – the Visible Human Project is only one of the more recent of its cybernetic innovations.

In consequence, biomedical practice has persistently and relentlessly thrown open the question of human limits and human nature, at the same moment that it asserts its natural stability. Hartouni (1997), for example, reports the now largely forgotten spate of anxious dystopian speculation around in-vitro fertilisation technics during the late 1960s and early 1970s, fear that the industrialisation of conception and birth would produce monstrous, trans-human entities. As Hartouni observes, this moment of anxiety has passed, and IVF technologies have been largely recuperated by a progressivist account of biomedicine.

> They are no longer regarded as contrary to the work of nature, but rather as instruments that promote or assist nature's work, enabling, correcting, or improving natural processes that have gone awry and that, in any event, apparently, are highly mercurial and inefficient.
>
> (Hartouni 1997: 114)

The intimation of the human as invention rather than inventor made available by IVF technologies has, she suggests, been thoroughly domesticated, naturalised and glossed over in the public imagination. The anxiety provoked by the prospect of human invention now seems to have transferred itself to the practice of cloning, as the recent successful cloning of mammals indicates that human cloning will become possible in the near future.

In a similar move the Visible Human Project provides a spectacular iconography of the human as invention. The Visible Human Project is, it seems to me, a spectre produced by the cybernetic turn, a data ghost which, in copying the human, theatricalises a very current moment of slippage in the category. As a double for the human, a medical mirror for species morphology, it fails to return a faithful, symmetrical image. Instead it displays the ease with which any idea of the human as given content is displaced, put into non-coincidental series through the fact of its reproduction. Moreover the VHP recapitulates earlier moments of slippage within the organisation of its apparatus – those thrown up by the development of anatomical practice at the beginning of scientific medicine, and by the development of the x-ray, moments discussed in Chapters 3 and 4 of this text. In what follows I want to consider the material and subjective implications of the posthuman moment inaugurated by the VHP. I will trace both the new kinds of relations it sets up between the terms of embodiment and technology, and the kinds of collisions it stages with the location and stability of the human subject as point of self-origin. As a data ghost it offers presentiments of a possible future, a spectre which allows speculation about the multiple ramifications of the cybernetic turn.

3

THEATRES OF VIOLENCE
The anatomical sacrifice and the anatomical trace

The Visible Human Project is an apparatus which recapitulates an entire history of anatomy within itself.[1] It is the most recent instantiation of a long biotechnical project, the anatomisation of human cadavers in order to produce the human body as a resource for 'Man', a technology at the disposal of conscious mastery. Through the dissection and analysis of the body's organ-isation, anatomy works to suspend any distinction between surface and depth, interior and exterior, endosoma and exosoma. It ideally makes all organs equally available to instrumental address and calibration, forms of engineering and assemblage with other machine complexes. Anatomical knowledge is, for example, a precondition for all internal surgical practice, which allows interior organs to be exteriorised and treated in linkage with life-support systems. In this way anatomy, supplemented with the knowledges of immunology, endocrinology and the like, intensifies the uses and capacities of organs, and sets out certain terms for exchanges between human and machine organs. In their earliest historical form these terms were a negotiation between the material trajectories of the dissected corpse and the anatomical text, the atlas which made possible the science of anatomy as such. Anatomy, like all biomedical science, has been quick to utilise new innovations in techniques of visual demonstration – photography, cinema, x-rays, tomography – and each innovation has demanded a new negotiation of terms, new practices of organ translation and exchange. The ramification of these terms is not limited to visualisation. They extend to the practices of surgery, organ transplant, donor cadavers, orthopaedics, prosthetics, diagnostics, and the whole panoply of medical biotechnology. The VHP apparatus effectively gathers up all of these moments of translation and exchange and reconfigures them within an economy of digital data.

The project also recapitulates a whole economy of biovalue which has underpinned anatomy from the time of its first exploratory moments in early modern Europe. Anatomy involves finding a use-value for the corpse, calling it to account in order to produce a surplus of vivification for the living. The corpse is anatomy's privileged object, and its foundational status in medical knowledge has decisive ramifications for ways in which life and death are distinguished in medical thought. Some of these ramifications are examined in Chapters 5 and 6, where I set out some of the conceptual slippages and spectral effects produced by medicine's utilisation of corpses as exemplary models of living anatomy. Moreover the Project revives and gives theatrical demonstration to another aspect of biovalue. In its use of an executed criminal as raw material for its enterprise, the VHP draws on a hierarchy within the human which Wolfe (1995), Gatens (1996) and other critics have demonstrated is structural to humanism as such.[2] Anatomy and other biomedical practices have, throughout the history of medicine, colluded with a more general sacrificial order which designates certain categories of people as more or less valuable than others, according to their closeness to nature and distance from the fully human. Anatomy and other biomedical disciplines frequently rely on the bodies of the less valuable and those deemed less human to act as model technologies or experimental objects, useful for the preservation of the health of the more fully human.[3]

In this chapter I examine these aspects of the VHP's prehistory. The hierarchical economy of biotechnical sacrifice and humanist redemption is discussed both in relation to the figure of Jernigan himself and the early history of scientific anatomy. This is followed by an account of the process whereby the human organ-ism is worked into a relation of exchange and translation with the organs of the book, the anatomical atlas which writes the earliest 'Book of Man'. The transformations of the body specific to bibliometric anatomy are contrasted with the transformations specific to virtual anatomy, in order to think the differential specificity of these media and the kinds of corporeal spatialisations they promote.

SACRIFICIAL ECONOMIES

In using the body of Joseph Jernigan, convict and murderer, executed for his crimes, as its first subject, the Visible Human Project refers back to a long history in which medical knowledge was intimately bound up with penal and sovereign power. From the fifteenth century until the mid-nineteenth century the barber-surgeons and academic anatomists of Europe used the bodies of executed criminals as the raw material for their investigations.[4] The anatomisation of criminals was frequently carried out in public demonstrations. Each important urban centre had its public anatomy theatre, and the dissection of criminals was a popular public spectacle, attended by the professional and fashionable. The anatomy theatres were kept supplied with corpses by successive pieces of legislation which placed the state's imprimatur on the use of criminal bodies, particularly the bodies of murderers, for the purpose. In Britain, for example, the Murder Act, 1752 sentenced the worst criminals to be 'hung, dissected and anatomised' (Forbes 1981).

This practice of public dissection was a spectacle which closely resembles the 'spectacle of the scaffold', Foucault's (1979) term for the torture, dismemberment and execution of criminals, carried out in a public square before the citizenry. Public dissections were frequently performed immediately after public executions, and constituted as extensions of punishment. Barker (1984) describes the annual public dissection of a criminal, held in the dead of winter before the burghers of Amsterdam, as a 'drama of retribution' in which the violence of medical dissection and the violence of punishment are indistinguishable, a fusion of punishment and science.

> Is not [the anatomist] as much the agent of an older punishment as the representative of a novel science? Does not the annual public dissection, conducted with solemn and awful ritual in the depth of winter, merely extend the execution which immediately preceded it and which provides its patient; and isn't this ... performance – which, at most, we might regard as some fusion of punishment and science – then completed by the commemorative banquet for the Guild of Surgeons by which it is followed? To execute, to dismember, to eat. It is difficult to imagine how much more than this an act of corporeal punishment might be.
> (Barker 1984: 73–4)

The Visible Human Project seems to present us now, at the beginning of the third millennium, with the return of the repressed; anatomy's historical dependence on, and resemblance to, the procedures of penal punishment and execution. The fate of Joseph Jernigan foregrounds this connection in dramatic terms. It draws our attention to a whole history in which the category of the human is established by excluding some from the status of lawful subject, and by simultaneously utilising their bodies as resources for the more fully human. Humanist discourse can, as Wolfe remarks, be strategically deployed against other human beings in the interests of establishing hierarchies. 'Humanism ... is species-specific in its logic (which rigorously separates human from nonhuman) but not in [all] its effects (such logic has historically been used to oppress both human and nonhuman others)' (Wolfe 1995: 36). Anatomy's iconography has, throughout the period of scientific modernity, been taken to signify the noble perfectability of 'Man' through knowledge and the harmony of rational order which he embodies. Only bring to mind Leonardo's famous anatomies, and their persistent deployment in popular culture to signify human progress, a signification recently reworked in the self-conscious juxtaposition of Da Vinci's anatomical work with the VHP figures themselves.[5] At the same time, anatomical knowledge has relied on the bodies of those excluded from the social contract to provide its raw material, those useless or dangerous beings at the margins of the human. While murderers remained the corpses of choice for British anatomists well into the nineteenth century, the Anatomy Acts of the 1830s also made the bodies of the indigent, the insane, prostitutes, suicides and orphans available for anatomisation (Richardson 1988). Here the splitting of 'Man' into species-being and subject described by Foucault (1975) is adjudicated into a hierarchy of beings: those who are designated as technical models of bodily order for the use of those designated as proper subjects. If the value of the former is marginal in their living state, their dead bodies can nevertheless be made useful, and their debt to the social order can be posthumously redeemed through the transformation of their flesh into knowledge.

This economy of human marginalisation, indebtedness and redemption is clearly expressed in the popular commentary on Jernigan and the VHP. An article about the project in *Life* magazine for example, describes Jernigan as a worthless murderer, whose family refused to pay for his burial, and who feared that he might be released from prison rather than executed. The article concludes that 'in death, he may finally do something good for humanity' (Dowling 1997: 41). Others describe him as 'a cruel and murderous drunk', a 'mad dog' and 'executed monster', literally a being beyond the pale of the human. Yet, as Visible Human, Jernigan is transfigured into a hero for human knowledge. An article in the *Chronicle of Higher Education*, for example, describes Jernigan as an 'internet angel'. 'In his life he took a life, in his death he may end up saving a few' (cited in Cartwright 1998: 32). As murderer, Jernigan steps outside the social contract, but as the raw material for the Visible Human Male he is understood to redeem himself by making a direct contribution to biovalue, preserving the life and integrity of bodies more valuable than his own.

It is instructive to note that this rhetoric of exclusion, debt and heroic redemption is absent in the case of the Visible Woman, an absence which is, I think, indicative of the different status of male and female bodies within humanist hierarchies, and within related hierarchies of useful matter. Jernigan's contribution to the formation of the VHP is acknowledged as having some causal status in the sense laid out in the previous chapter; *causa* – a being-indebted-to. While on the one hand the criminal subject, Jernigan, was in debt to the social order from which he had stolen a life, on the other this debt is redeemed by his heroic gesture in donating his body as useful matter, giving his form to the formation of the project. In popular culture at least, Jernigan's contribution to the VHP, to putting its iconography in train, seems to be widely acknowledged and celebrated.[6] Meanwhile the Visible Woman's contribution seems to be taken for granted as the donation of one always/already marked as useful yet passive matter, and as reproductive matter already belonging in some sense to the community. This evaluation is clarified further by the deployment of Jernigan's entire body as exemplary text, while the VHF is most readily available as a set of reproductive organs. While the Visible Human Project might be the most recent form of anatomical imaging, it nevertheless recapitulates a long history within anatomy and medicine more generally where the male body is considered the exemplar of the human body, while the female body is treated as a reproductive adjunct. As Moore and Clarke (2000) demonstrate in their survey of various kinds of cyberanatomies, the application of the latest technology to anatomical imaging does not in itself produce more progressive imaging of sexual difference.

Anatomy has historically relied upon the bodies of the socially marginal as its raw material because of the spectacle of violence which subtends its most benign iconography. The anatomical image can never be simply a benign reflection of the human body as referent, an illustration of its pre-existing anatomical order. Its conditions of production – the way it calls the body to account – involve the sacrificial dismemberment of particular bodies in the interests of textual knowledge production. In this regard it is continuous with the entire field of the biosciences, which always involve some degree of violence, in the sense of a forceful technical intervention in the organism. As Keller (1996) indicates, the generation of bioscientific knowledge always requires some degree of 'literal, material transgression' of the object to be understood. While biomedical and biological knowledge has been conceptualised in cultural studies as primarily scopophilic, organised through the logic of the gaze and visualisation, she points out that this gaze is always projectile, associated with a transforming and interventionist touch, with 'taking the object into hand, ... trespassing on ... the very thing we look at' (Keller 1996: 108). Like all biological knowledge, medical knowledge is produced through an analysis of the flesh which involves some technical reordering of the flesh as its condition.

The anatomical dissection takes this relationship to its logical conclusion, involving as it does a thorough disaggregation of the body into its organ-parts. To anatomise is to analyse, to partition, to reduce a whole to its constituent parts. In particular the anatomical dissection is concerned to bring the concealed interior of the body to light, rendering opaque depth as visible surface. Each layer of flesh, sinew, muscle and nerve is peeled away, each organ separated out so that it can be presented to the eye of the anatomist, made visualisable. In this sense anatomy is one of the most graphic demonstrations of the body treated as organ-ism, as an assemblage of tools, whose value lies in its capacity as useful machine.

Hence anatomy involves the sacrifice of one body as a model technology, in order to place its workings at the disposal of others. The anatomical dissection can be considered as a sacrificial practice in a number of senses. First, it requires a death before it can take place. 'Sacrifice' is, in fact, a technical term within the biosciences for the killing of laboratory animals, 'in preparation for, during, or subsequent to their use as experimental subjects' (Lynch 1988: 265). While many other biotechnical knowledge procedures can be performed on abstracted body fragments (e.g. cell lines) or through instrumentalising a living body (e.g. clinical trials), anatomy can only legitimately dissect a dead body, and anatomy as a discipline is based on an accumulated knowledge of corpses. While anatomy no longer depends on punitive execution or forceful appropriation of dead bodies for its raw materials, Jernigan's trajectory from incarceration to execution to anatomisation retains a trace of such enforced sacrifice, a compulsion not entirely dispelled by the fact of his consent to a request by prison authorities to donate his body to science. While the project directors did not set out to use the body of an executed criminal, the choice of his body, as opposed to other available bodies, was, as Cartwright (1997) argues, directly determined by his status as penal non-citizen and object of institutional violence.

> The Colorado team that selected the body for the Visible Human Project claims not to have learned of the identity and source of Jernigan's body until after its selection. ... [However] Jernigan's body would never have come up for review by the team had Jernigan not been an anonymous and marketable cadaver, a subject whose status as live ward of the state rendered him in death a subject stripped of his rights to certain privileges of citizenry – most notably, the right of bodily privacy. ... It is precisely because of Jernigan's status as less than private citizen, as a subject stripped of certain rights under the auspices of the state, that he qualifies as universal biomedical subject, and as a public icon of physical health.
> (Cartwright 1997: 136)

In the woman's case, the more conventional 'donation' of her body to science makes her gesture a 'self-sacrifice', a posthumous gift of the body to knowledge, although an unconfirmed rumour claims that her body was specifically donated to the Visible Human Project by her husband.

Sacrifice does not simply involve death. It also involves transformation. As Lynch writes of his study of animal sacrifice in the laboratory,

> 'Sacrifice' ordinarily implies an act of 'making sacred' in terms of a western dualism between 'sacred' and 'profane' realms. While the mundane laboratory animal is not transformed into a 'sacred' object *per se*, its material body and the interpretive sense of that body are radically transformed through a series of preparatory practices which turn the animal into the bearer of a generalised knowledge.
> (Lynch 1988: 266)

The sacrifice of organic entities, their progressive dismemberment, is performed in order to transform them into data, to render knowledge gained from the interiorised processes of life as graphs, indices, preserved body fragments, micrographs – an endless list of documentary or residual forms. In the case of the anatomical sacrifice the corpse is dismembered in order to transform it into particular kinds of images – images of the ordering of the bodily interior which retain a mathematics of proportion and spatial orientation with regard to an imagined whole body. To paraphrase Lynch, the anatomical process transforms everyday corpses, locatable entities in everyday space, into *analytic* entities, which take place in a 'generalised mathematical space', and which are endowed with paradigmatic status. They are claimed to be able to stand for human bodies in general, and to be applicable as precision maps of bodies in general. One of the unique features of the VHP figures is that, unlike conventional anatomical images, they cannot lay claim to anonymity. In general, the transformation of corpses into analytic entities in conventional anatomy texts effaces particularity. Anatomical images are usually summary images, produced by synthesising the results of several dissections and normalising these results so that they can act as exemplars of an ideal, healthy morphology (Daston and Galison 1992). While the production of the conventional anatomical text demands the sacrifice of particular bodies, the images produced are not particular to any once living person. In the case of the VHP however the transformation from an everyday to an analytic entity has preserved the individuality of each figure in a kind of anatomical portrait, a personal photograph. Despite the dramatic violence of their process of production, the distinction between the everyday entity, the particularity of the once living person, and the analytic entity, the anatomical text, seems weak, and the transformation incomplete. This accounts, in part at least, for the poignancy of the images, their tragic appeal and their uncanny air of absent presence, of identity lost yet not erased.

Finally, anatomical processes are sacrificial in that they imply a double structure of redemption. As I already pointed out, the anatomical process can, even today, lend itself to a moralising discourse, in which a *mauvaise sujet*[7] can redeem his lost human status through productive death. In doing so he does not simply redeem himself; rather, like any sacrifice, his death and dismemberment is carried out in order to save the lives of many. Sawday brings together all of these moments of sacrifice in a single observation.

> Dissection is an insistence on the partition of something (or someone) which (or who) hitherto possessed their own unique organic integrity. But dissection or anatomisation is ... an act whereby something can also be constructed, or given a concrete presence. In medicine anatomisation takes place so that, in lieu of a formally complete 'body', a new 'body' of knowledge and understanding can be created. As the physical body is fragmented, so the body of understanding is held to be shaped and formed. In medicine, too, anatomisation takes place in order that the integrity and health of other bodies can be preserved. The anatomist, then, is the person who has reduced one body in order to understand its morphology, and thus to preserve morphology at a later date, in other bodies, elsewhere.
> (Sawday 1995: 2)

Hence the anatomical image is produced through a process of dis-integrating the corpse, and its status is closer to a material residue than to a mimetic illustration of the body. Illustration connotes a derivative procedure wherein the coherence of an object is reproduced as a mirrored content in some medium, pencil on page or light emulsion on photo-media. In the terms of illustration, both entities remain in a relation of self-contained coherence and hierarchical reflection – one simply records the other. The anatomical image however is produced through a process which literally writes flesh on the anatomical slab according to certain logics which enable it to be transformed into a structure of iteration, a system of legible and repeatable inscriptions, the image-writing of the anatomical atlas. Anatomisation, by which I mean both the process of dissection and the production of anatomical atlases, is, in effect, a procedure for transforming the volumetric bulk of the singular, locatable corpse into a readable, and hence writeable, object. Anatomisation could be thought of as a series of spatialising practices which successively transform the bulk of the three-dimensional body into iterable traces, a system of image-objects which map the body into media space and enable the replication, circulation and accumulation of anatomical knowledge. The protocols of dissection, which I will investigate below, are not dictated by some inherent logic of anatomical organisation, some pre-existing organic order, but by the necessity to cut the body up along lines which lend it to the spatial ordering of particular media, the book or the computer, as material objects with particular technical trajectories.

The development of the VHP has foregrounded the extent to which anatomy's methods of writing the body are, like all forms of writing, *medium specific*. That is to say, the technical conditions for the reproduction of traces, the writing media, condition the ways that anatomical meaning can perform and circulate. As I will discuss, the materiality of the book and the computer enter directly into the materiality of the anatomised corpse in this process, laying out the logics according to which it is to be written. The violence of anatomy is the violence of a particular kind of textual practice, one which demands the sacrifice of that which it writes, for which death and dismemberment is a structural condition of the production of the trace. This is the pathos of anatomical images. They demonstrate the body's workings, its vitality, yet they also necessarily allude to the body's mortality, its ineluctable drive towards absence, and its vulnerability to the violence of biomedical imperatives. At the same time this rendering absent is the precondition for the atlas's efficacy as biomedical prosthesis. By finding techniques for reading and writing corpses according to the enframing logics of particular media, anatomy has helped produced ways of therapeutically enframing *living* bodies, re-spatialising them through the practices of surgery. It is this writing of the living, the tele-surgical, which will be taken up in the next chapter.

In what follows I want to consider the emergence of the earliest scientific anatomical atlases and analyse their corporeal textualisations in relation to those practiced in the VHP. The purpose of doing so is to address what

Weber (1996) terms the *differential specificity* of the medium, to think about the Visible Human Project not as an essential anatomical content expressed in a new medium but as a transformation of the material possibilities of the anatomical body inaugurated by a move from the book to the screen. This is not to undertake an historical argument per se, in the sense of providing a systematic, chronological account of the VHP's precursors and their social context. Rather, by comparing the moment of the emergence of the anatomical text with that of the virtual anatomy I want to demonstrate the ways in which the media of anatomical demonstration condition anatomical knowledge and practice, and in turn lay out the material conditions for the formation and circulation of normative human archives. By adopting a comparative method I am following Weber's (1996) caution against ontologising the qualities of any medium. He writes 'The attempt to work out the *differential specificity* of [a] medium – to get at that which distinguishes it from other media – runs the risk of transforming ... a differential determination into a positive and universal essence' (Weber 1996: 109).

FROM CORPSE TO ARCHIVE

The practice of dissecting corpses anatomically marks the historical threshold between pre-scientific and scientific medicine, the moment, sometime in the early sixteenth century, when European medicine ceased to be a practice based on the received ancient wisdom of Galenic medicine, and began instead to be based on empirical practices of observation, a testing of classical knowledge against the experience of the body.[8] Galenic anatomical practices generally left human bodies intact, part of an older theological order which regarded the interior of the body as sacred, the domain of the soul. Its anatomical images were produced and transmitted by hand-copying images in older texts, without being rerouted through an actual encounter with material bodies. Modern medicine is conventionally understood to have begun when the regular dissection of corpses became integrated into its knowledge procedures.

The *Theatrum Anatomicum* of the late Renaissance was, according to Sawday, an architectonic machine through which the orderly structure of the human body and its orderly position within the cosmos could be laid out before an audience of witnesses. The anatomy theatre's optical arrangements and ratios were Vitruvian,[9] designed to place the body at the central point of a circular structure which both permitted a maximum and evenly distributed visibility and emphasised the rationality of the relations of part to whole, both in the body itself and between body and space. This space of rational optics dramatised the anatomical process as an exemplary spectacle of the new scientific knowledge. The *Theatrum Anatomicum* was a space where the new commitments to empirical knowledge procedures and a rational knowledge of 'Man' could be enacted. Hence the confrontation between anatomist and corpse in the space of the anatomical theatre

> expressed a view of knowledge itself which was at a point of metamorphosis: for the anatomist who searched in the body for its structure rather than in the texts of ancient authority, was the concrete representative of a new conception of knowledge, one that professed to rely on the experience of phenomena rather than the experience of textual authority. ... [Furthermore] the anatomist, in his scientific jurisdiction over and above the criminal body, expressed the symbolic power of knowledge over the individual, a continuation of the process by which the individual was forced, on the gallows, to acknowledge the legitimacy of the sovereign power over his or her body.
>
> (Sawday 1995: 64)

The anatomical theatre is, then, a technology for staging displays of rational mastery over the dead body of the criminal, mastered not only as a being to be punished but also as a material object to be explicated according to the dictates of its own rational order.

Dissection in itself was not sufficient to establish the set of knowledges systematised as human anatomy, however. The possibility of an archive of the human body, an exhaustive, empirically verified, serial compendia of positive informational contents, began with the coming of the mechanically printed book. While the anatomical demonstration in the *Theatrum Anatomicum* produces an immediacy of knowledge about the human body, it is, by definition, a mercurial and unstable knowledge, a time-based effect which dissipates after each particular performance. A corpse is transient, rapidly corrupting, unable to be preserved for any length of time. It can be inspected by the anatomist and demonstrated to an audience, but the experience of sight is itself fleeting. Any *cumulative* knowledge of the corpse only becomes possible with the development of processes of recording. As one historian puts it,

> Anatomy as a science ... was simply not possible without a method of preserving observations in graphic records, complete and accurate in three dimensions. In the absence of such records even the best observation was lost because it was not possible to check it against others and thus to test its general validity ... in the observational or descriptive sciences illustration is not so much the elucidation of a statement as a statement itself.
>
> (Panofsky, cited in Eisenstein 1979: 268–9)

Anatomisation involves the production of bio-graphs, trace systems which demonstrate the ordering of organs and tissue in relation to the whole body as it is oriented in Euclidian space. This bio-graphic yield is the result of the sacrificial transformation noted earlier, the dismembering of organic entities in the interest of trace production. This trace production is, on Latour's (1990) account, the object of scientific work. Science is precisely a method of *writing* natural objects, of working them over in order to produce a repeatable material residue (an x-ray, a spectrograph) or summary figure (graphs, tables, formulae), trace systems which refine and display the qualities of interest for the scientist, and which lend themselves to what Latour terms 'capitalisation' – the trend towards simpler and simpler inscriptions which mobilise more and more information about the object, and propagate it through more complex forms of circulation, interpretation and communication within and beyond scientific communities (Latour 1990: 41). Capitalisation in this sense is a central mechanism in the production of biovalue, a way to potentiate the knowledge gained from any systematic, empirical encounter with organic entities and increase its yield of biotechnological effects.

Hence the centrality of the printed book in the formation of scientific knowledge generally, and anatomical knowledge in particular. Mechanical printing provided a method of preservation, circulation and reproduction of the knowledge gained from particular dissections, a way to move that knowledge away from local conditions of time-based performance and propagate it through networks of readership. The book, Latour writes, particularly the scientific atlas and the cartographic atlas, is a mechanism for mobilising time and place. Printed books are able to accumulate and disseminate knowledge about other locations in immutable, mobile inscriptions – drawings, maps, graphs, diagrams and the like, which can demonstrate what they describe.

> For the first time, a location can accumulate other places far away in space and time, and present them synoptically to the eye; better still, this synoptic presentation, once reworked, amended, or disrupted, can be spread with no modification to other places and made available to other times.
>
> (Latour 1990: 32)

Prior to the invention of the printing press, the technical conditions of medieval scribal culture were antithetical to the development of visual forms of knowledge, precisely because of limitations on the iterability and circulation of traces. Scribal illustration and copying could only be carried out on a very small scale with very limited circulation, and was subject to constant degradation of the original image, a constant loss of accuracy and quality. Under these conditions intellectual discoveries and observations could be, and were made, but,

> each achievement stayed local and temporary just because there was no way

to move their results elsewhere and to bring in those of others without new corruptions or errors being introduced. For instance, each carefully amended version of an old author was, after a few copies, again adulterated. No irreversible gains could be made, and so no large-scale, long-term capitalisation was possible.

(Latour 1990: 34)

By contrast to this situation of image-entropy, Latour describes the printing press as a means of preserving and improving on visual knowledge, a mechanism 'to irreversibly capture accuracy' in the reproduction of images, a mechanism for the production of reliable, demonstrative, visual knowledge.

> Engineering, botany, architecture, mathematics, none of these sciences can describe what they talk about with texts alone; they need to show the things. But this showing, so essential to convince, was utterly impossible before the invention of 'graven images'. A text could be [hand] copied with only some adulteration, but not so a diagram, an anatomical plate, or a map.

(Latour 1990: 34)

The anatomical atlas is a kind of book dedicated to the documentation of the anatomical dissection. On Latour's analysis the anatomical text is a way of delocalising the dissection, rendering its performance in an immutable, mobile form through a maximisation of visual display on the page and a presentation of visualised organ-objects with minimal forms of commentary. Like the dissection itself, the atlas lays the corpse out before an audience, presenting its interior to the eye through an orderly display on the page. Nevertheless, the atlas does not simply extend the reach of an anatomical performance as a documentary adjunct, a recording of the body's organic logic as it is revealed in the dissection. Rather, it is, in itself a spatial technology which largely drives the form of the dissection, ordering the corpse in accordance with the spatial logics and material trajectories of the atlas itself.

What I mean by this is that there is nothing straightforward or transparent about the way bodies can be dissected and turned into atlas illustrations. Sawday (1995) describes a long interregnum in which the early anatomists, faced with the wet, bloody mess which is the corpse's interior, were at a loss as to how to order it. Dissection clearly involved opening up the body, turning its self-enclosed interior to the exterior and the light, but the ways to open it up were not self-evident. He relates attempts to dissect bodies by working from head to foot, and by slicing the body down the middle, 'opening it as though it were a book, whose spine was (literally) the human spine' (Sawday 1995: 132). *The systematisation of dissection was a process wherein the human body had to be rethought in relation to the body of the book, the forms of its material extension, its particular operative modes and spatial capacities.*

The anatomical body is co-emergent with the anatomical text, and only begins to take shape when the organ-ism is itself addressed during the dissection as a form of book – as a readable terrain, whose telling surfaces can be transcribed cartographically, laid out serially in the serial space of the book, and registered spatially in the projected Euclidean space of the page. So the human body is made into an archive whose content cannot be disengaged from the material form of that which purports to simply contain it. Like the later HGP and VHP archives, this earliest anatomical archive, the anatomy atlas, presents an analysis of the putatively natural human morphology by reworking it according to the technical logics of the archive itself.

While this is not an exhaustive account, there seems to me to be two crucial epistemological moves through which this transformation was achieved.

1 THE BODY AS LAMINAR TERRAIN

The central problem of anatomy is the incommensurability between the opaque volume of the body and the flat, clean surface of the page. By what means could this volumetric bulk be transposed onto this two-dimensional surface? This problem demanded the consideration of how material objects related to the space of the page, a problem which was general to European Renaissance culture, and which was resolved to a large extent through the invention of perspective drawing (Latour 1990). While the deployment of perspective was crucial in the production of anatomical drawing, it did not address the problem of how to turn the volumetric thickness of the body into visualisable material.

This problem was resolved to some extent through the creation of analogies between anatomical and cartographic space, analogies evident in the fact that the book of anatomy is known as an atlas. If the interior of the body could be thought of and treated as *space*, rather than as a self-enclosed and continuous, solid volume, then it could be laid out in ways which are amenable to a form of mapping, just as the navigators of the time were mapping the new world. In this analogy the body's interior was described as a material terrain to be surveyed, a new topography which demanded exploration and discovery.

> The task of the scientist was to voyage within the body in order to force it to reveal its secrets. Once uncovered, the body-landscape could be harnessed to the service of its owner. In thus establishing the body as 'useful' … we are able to perceive the language of colonialism and the language of science as meshing with one another.
>
> (Sawday 1995: 25)

As Sawday (1995) comments at numerous points, the visual style of the Renaissance anatomical atlas and the Renaissance map shared much in common, both laying out newly charted and named territory using newly developed conventions like keys, which allowed narrative information to be co-ordinated with visual information, while preserving the centrality of the latter. By conceptualising the body as uncharted terrain the anatomists were also finding a mode of legitimation for their enterprise. If the body was already a kind of space, then its traversal seems less violent, and if it is an uncharted space, then it is the responsibility of the anatomists and scientists to chart it, to make it intelligible in order to make it a useful resource, rather than an opaque danger, a site of unknowable pathologies. To write the body as anatomical space is to mark it as a space controlled by medical logic and under medical fiat, rather than other possible systems of meaning.

This analogy between corporeal and cartographic space simplified the problem of how to translate the three-dimensional volume of the body onto the space of the page, because it conceptualised the body as an accretion of laminar 'surfaces', as landscapes to be traversed by the eye, a volume composed of layers and systems of tissue which are laid one upon the other. In the dictates of pathological anatomy, Foucault writes, it was crucial to 'break down ... these organic masses into tissular surfaces ... if one is to understand the complexity of function and alteration' (Foucault 1975: 128). Hence the anatomical demonstration proceeded, ideally at least, as a *flaying*. The *échorché* figures of classical anatomy are produced through the peeling away of first skin, then successive layers of muscle, to work through the body's volume as if through the layers of an onion, from outer surface to bone. (See Figure 3.1.) As facialised surfaces these anatomical features could then be mapped in succession.

Clearly this graduated abolition of depth is continuous with the necessity to render the body demonstrable in a visual mode. This topographical or laminar spatialisation of the body was also precipitated by the spatiality specific to the book, another laminar volume. The body's volume could be laid out sequentially in the book's volume, and the reader could traverse the former by flipping through the latter. Furthermore, the form of the book suggests *both* a spatiality and a temporality, in so far as one reads and turns its pages in a sequence. Hence this form lends itself to a spatialised narrative about the body's constitution, an ordering which reverses the ideal spatial order in which the corpse is dissected. Vesalius in *De Humani Corporis Fabrica* (1543)

> deployed an architectural mode of analysis in that it envisaged the body as 'constructed', and it sought to replicate this construction by gradually building up the various detailed segments into an organised whole. So, the Vesalian text began with the skeleton, and then considered the muscles,

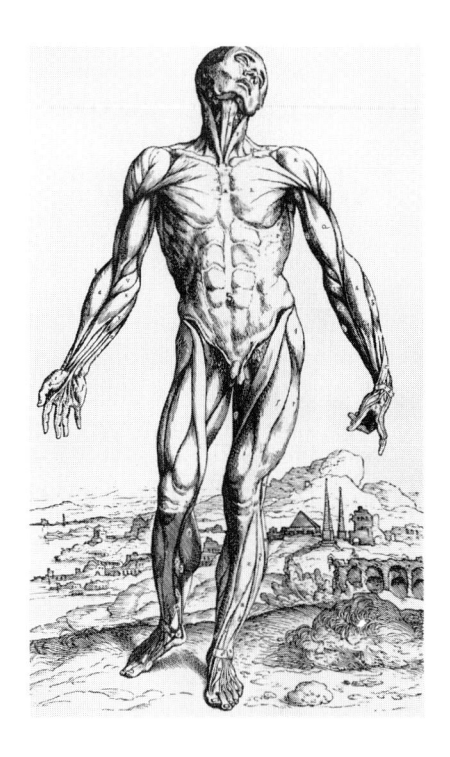

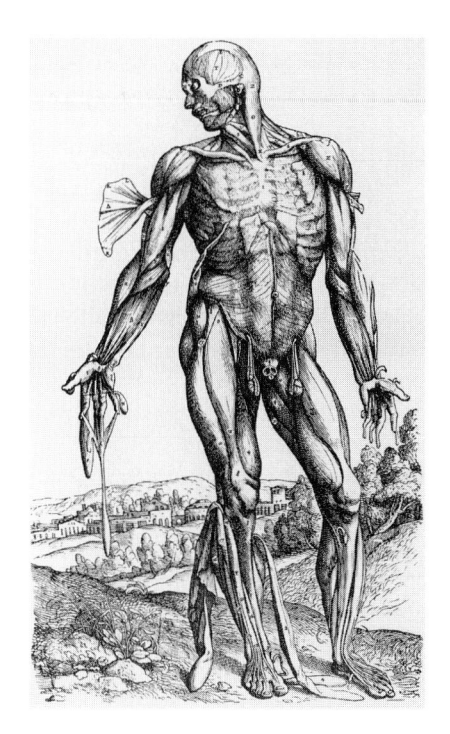

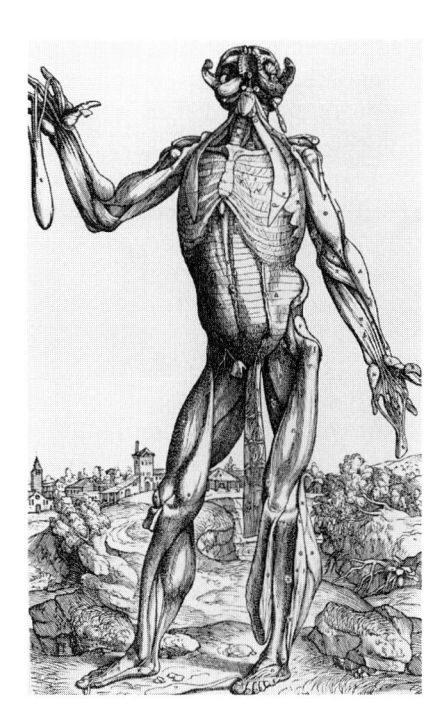

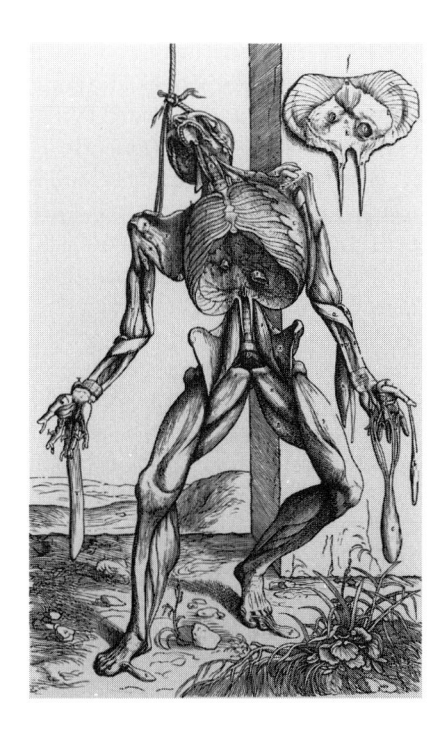

Figure 3.1 Plates showing some of the muscles, from Andreas Vesalius' *De Humani Corporis Fabrica*, 1543

the vascular system, the nervous system, the organs of nutrition and the abdominal viscera, to end with the brain.

(Sawday 1995: 132)

By narrating the body as a sequence of systems and an assemblage of organs the anatomical atlas creates the sense of a body which is always/already only the sum of its essentially discreet parts. These parts may act in co-ordination in the living body, it implies, but they have a pre-existing separateness which allows them to be isolated and identified in the act of dissection. According to the logic of the anatomical atlas the volumetric self-enclosure of the body is only ever provisional, and somewhat incidental to its true organisation.

2 THE BODY AS READABLE TERRAIN

The practices of anatomisation work on the assumption that the interior of the body is a legible surface, which presents intelligible traces to the eye of the anatomist. As numerous commentators (Stafford 1993, Jordanova 1989) have pointed out, the inauguration of scientific dissection and the systematic production of atlases are important moments in the general scientific privileging of vision as the appropriate sense for the apprehension of rational knowledge. Harcourt (1987) argues that the reason that Vesalius' volumes of anatomical text, *De Humani Corporis Fabrica* (1543), are conventionally celebrated as the moment when modern medicine was inaugurated, is because 'it established anatomy once and for all as a discipline absolutely dependant on a [visual] system' (Harcourt 1987: 53). The anatomical dissection treats the body according to the demands of a science which, as Foucault (1975) puts it, is ordered on the 'exercise and decisions of the gaze', around the logics of visual identification and recognition of bodily organisation. Rather than utilising the explanations of heat, cold, wetness and dryness deployed in medieval humoural medicine for example, anatomy was premised on the discernment of visible characteristics, forms, structures, surfaces and lesions. The anatomical gaze demanded 'to see, to isolate features, to recognise those that are identical and those that are different ... to grasp colours, variations, tiny anomalies' (Foucault 1975: 89). Pathology, too, is conceived on a model of visualisable space; 'Disease ... is a set of forms and deformations, figures and accidents and of displaced, destroyed, or modified elements bound together in sequence according to a geography that can be followed step by step' (Foucault 1975: 136). Certain Enlightenment clinicians developed a whole system of 'reading' pathologies, of seeking an 'alphabetic structure of disease', laws of combination of visible symptoms which allowed the language of illness to be deciphered like a word or sentence (Foucault 1975: 118).

Here the interior of the body is itself readable as a system of discernable traces, a pathographic meaning system which takes place in the same kind of hermeneutic space as does the book itself. It is unsurprising then that for the early anatomists the act of anatomisation was considered an act of reading, and the body was treated explicitly as a kind of book, whose coherence and ultimate intelligibility derived from the coherent intentions of its divine author which can be discerned through a faithful interpretation. The anatomical body is

> constituted as text – as the *liber corporum* – the book of the body written by God. ... The anatomist, then, who 'reads' the anatomy (relating body and text together in the anatomy theatre) was reading two different kinds of text: a text written by human agency (the observations of his predecessors) and a text written by God, comprised of all the different members, sections, subsections, and partitions revealed in dissection. The task was to recreate, in order to read, the precise system of division by which the body-book had originally been composed by the divine author.
>
> (Sawday 1995: 135)

While Sawday locates the *liber corporum* as an entity specific to Renaissance anatomy, there is evidence of this locution much later in the history of anatomy. Marvin (1994) relates the habitual description, during the eighteenth century, of bodies in the dissection rooms of the London Royal College of Surgeons as 'books', from which neophyte surgeons should learn. Logically enough this reciprocity between book and body also produced the occasional instance of 'anthropo-dermic bibliopegy', the binding of books in human skin, a practice favoured by nineteenth-century medical men for the binding of anatomy texts in particular.[10] The currency of the *liber corporum* is an issue for more detailed historical research and debate than is possible here, but it seems plausible that a direct if occasionally interrupted line could be drawn connecting this earliest usage with that of the contemporary 'book of Man', the Human Genome Project.

If the body is addressed in the dissection as itself a book, then the anatomical text, used as an aide in dissection, acts as a secondary commentary, a primer which enacts a visual exegesis on the primary text, the corpse. It performs the work of simplification, analysis and guidance. It alerts the reader of the body to functional relations, interlocking systems, and to distinctions between the important and unimportant within the confusing mess of the corporeal interior, eliminating what it considers superfluous detail in order to maximise intelligibility. It is a tutelary text which educates the eye and hand of the anatomist, literally teaching him or her to read in the language of anatomy. David Armstrong, writing of his own experience as a medical student, describes this tutelage in the following terms:

> The anatomical atlas directs attention to certain structures, certain similarities, certain systems, and not others, and in so doing forms a set of rules

for reading the body and making it intelligible. In this sense the reality of the body is only established by the observing eye that reads it. The atlas enables the anatomy student, when faced with the undifferentiated amorphous mass of the body, to see certain things and ignore others.

(Armstrong 1983: 2)

Hence the anatomical atlas developed a repertoire of graphic conventions which schematise the body's interior complexity as it is made to appear on the page. Anatomical texts habitually utilise a certain clean aesthetic which involves the hard-edged, crisp delineation of organs and tissue, the designation of particular functional systems through colour coding, the isolation of particular body parts in the space of the page, and the exclusion of blood or other body fluids from the image. (See Figure 3.2.)

They map the body interior as legible system, a system in which each organ and layer of tissues is aesthetically specified as a functional, discrete component in a systematic machine. If the anatomical text instructs the reader in the legible language of the body, it is a machine language, able by implication to be addressed and utilised by other instruments, other machines.

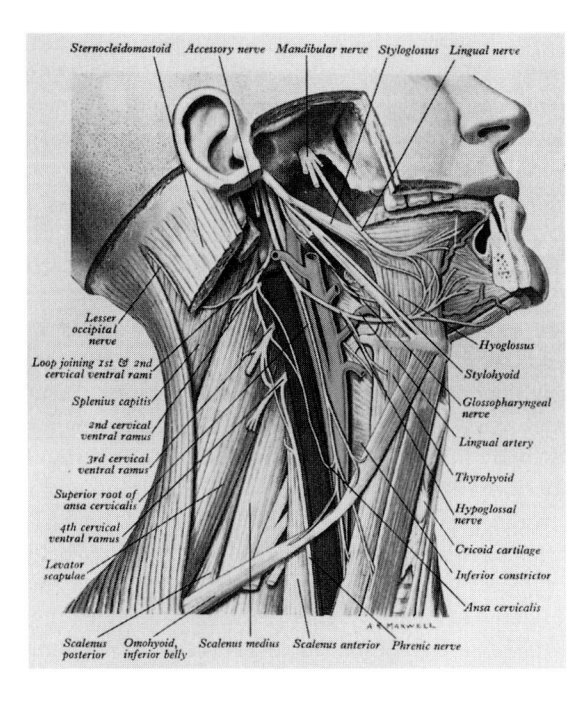

Figure 3.2 Dissection showing the hyperglossal and lingual nerves, in *Gray's Anatomy*

In the practice of internal surgery, a practice anticipated by the anatomical text by several hundred years, the interior of the living body is therapeutically addressed as an assemblage of (in this case) malfunctioning technologies – organs to be fixed or replaced by other working parts. The atlas acts as a guide in this process, but it is only useful in so far as the bodily interior is reworked according to the atlas's logic, a surgical process termed 'exposition', a setting out or legible *explanation* of the patient's body. It is only after this work of explanation that the body can be effectively treated as operable organ-ism. Hirschauer describes the work of exposition in his ethnography of surgical practice.

> Dissection, which is the precision work of making objects visible, is at the same time classifying work. The flesh is dense and compact, stuck together and impenetrable. First, one has to identify something in a crevice opening up, in the depths of a wound or on a bloody surface. ... In the case of microsurgery, this identifying work can take hours, in which whitish and reddish cords are identified as particular nerves and vessels and lifted out of their bloody surroundings by slings and numbered clamps. ... Dissection aims to present organs in the isolating style of the anatomical atlas. The drawings show neatly separate organs; in the patient-body this state must first be produced by isolating them with the knife. Surgeons call this 'exposition' or 'making anatomy'. ... When the exposition is completed, the target organ can be operated on: nerves are anastomosed, prostheses implanted, organs resected, tumours extirpated, bones screwed together.
> (Hirschauer 1991: 300–1)

It seems then that anatomical hermeneutics demand a simultaneous reading and writing of the body, a co-ordination of eye and hand in a circular economy of interpretation and inscription.

Through these techniques of laminar spatialisation and corporeal *explanation*, and doubtless other techniques as well, the earliest scientific anatomies established the technical logic of the human archive. *De Humani Corporis Fabrica* presents an exhaustive visual tour through the human body, laying its bulk out as a succession of systems, from the most superficial – the dermal and myological – to the most profound – the skeletal. In doing so it makes a Euclidean map of a standardised human body, specifying the spatial interrelationship between parts, and between part and whole. It also makes an inventory of the body as a collection of useful organs, an organ-ism whose functional parts can be potentially instrumentalised as use values for the subject. And it makes these features accessible and retrievable through indexes, systematic pagination and narrative ordering.

While the grammar of the book's title, 'On the Structure of the Human Body', indicates that these images are to be considered as an essential content, which the book merely contains as a transparent medium, nevertheless the atlas hopelessly confuses container and contained. By arraying the body's parts throughout the serial array of its pages as sequences of flesh, *De Humani*, and any other systematic anatomical text, implicate the volume of flesh in the volume of text, assimilating each volume to the other. The space particular to the book and the page, the space of bibliometric volume and Euclidean geometry, the mathematics which projects a propositional three-dimensional space on a two-dimensional page, sets out the orientation points for the systematic sacrifice of the corpse in the anatomisation. As Lynch observes, laboratory sacrifice, the particular protocols through which experimental animals are killed and rendered, always takes place with reference to 'an ideal set of spatial coordinates, a three-dimensional matrix that pervades the rendering process and acts as the basis for the integration of the animal's remains with a graphic display of data' (Lynch 1988: 277). Without such systematic orientations the sacrifice is merely dead, and its utility for knowledge production is wasted.

In this sense not just the inscriptions on the page but the whole volume of the atlas is a kind of trace, a material residue of the corpses sacrificed and dis-integrated in its process of production. In textualising corpses as anatomical volumes the anatomisation does not simply assimilate one body to another however. Rather, it sets up a bilateral relationship between the two. It is because of the spatial precision of the anatomical sacrifice – its careful orientation to the spatiality of the atlas – that the atlas can be used as a set of operative images which assist the surgeon to leaf through the living body of the patient, to anticipate structural relationships within its opaque depth, that is *to operate* within the bodily interior as intelligible space. This operation is discussed at length in the following chapter.

To contend that anatomy is medium specific is to consider the constitutive part played by the materiality of the anatomical trace and its particular modes of reproduction and circulation. I have argued that the coming of the book allowed modern medical knowledge to take form, but that it took form in a particular way, the form enabled by the particular conditions of possibility presented by the book. The book determined what kind of trace of the material body could be deployed in medical knowledge, how these traces were to stand to that body, and what problems and opportunities of interpretation, accumulation and circulation they presented in the complex task of moving from body to book and back again. These conditions are radically transformed by the advent of virtual anatomy.

VIRTUAL ANATOMY: FROM MAP TO SURROGATE

This early history of anatomy is thoroughly inscribed and replayed in the Visible Human Project. I have already commented on its recapitulation of the relations between punishment and scientific sacrifice. The project also recapitulates the operations of the *Theatrum Anatomicum*, which redistribute themselves throughout the networks of the internet. The VHP is by no means the only or the first anatomical site on the Web. On the contrary, there are, at time of writing, more than 100 anatomical imaging sites listed on the web, and an abundance of anatomical databases which can be searched with Web-based technology. Nevertheless, the launch of the Visible Human Male, and its attendant media fanfare, has solicited a general audience for this new theatre of anatomy which was not engaged by the more specialised services of the anatomical databases. This Web-based theatre enables the user at the workstation to sample images from the project, to play animation flythroughs, and with the right software, to interact with the data, performing limited kinds of virtual dissections. The properly equipped and licensed user can download the entire database and work the data themselves. Hence the operation of the *theatron* as a technology for facilitating sight, and for resiting objects before the audience is propagated not through an architectonic space but through the numberless networks and virtual sites which are the Web. The delocalisation of the traditional anatomical theatre is accompanied by a related movement whereby each terminal becomes a theatre *in potentia*, a place from which to see the Visible Humans' dissection, not as a single event to be witnessed in real time but a simulated, reversible, and endlessly repeatable event.[11]

Like the book-based atlas the VHP also reads and inscribes the bodily interior as intelligible visual trace, amenable to the reproduction, circulation and accumulation possible within the economies of particular media. Like the book-based atlas, the VHP is a technology for writing bodily space; its difference from the conventional atlas is that it writes not as analogue inscription but as digital data. It reads and writes the anatomical body according to the logic of the screen and the computer rather than the page and the book, according to the new modes of visualisation, distribution, replication and collaboration made possible by the computer and computer-mediated communication (CMC). In doing so it produces a visual archive of the human whose performative qualities are quite different from and more dramatic than those provided by the book. I will first describe some of these performative qualities, and then return to the question of the sacrifice which underwrites the production of the anatomical trace.

1 *VOLUMETRICITY*

Much of the appeal of the VHP for medicine and the biotechnology industries resides in the complex form of spatial 'capture' and modelling that virtual space is able to sustain. In this regard the VHP represents a solution to the problem of the awkward incommensurability between bodily bulk and flat page discussed in relation to the atlas. The space of the page and the book renders the volume of the body by transposing its complex contours onto a flat surface through visual conventions of perspective which indicate the way this transposition is to be interpreted. The book-based atlas necessarily sets out the relationship between the body's inside and its outside as a sequence of surfaces, successively laid out in the spatial and temporal order of the book.

By comparison, the VHP is able to register volumetric space with much more complexity. Its tomographic mode of production – the planning, photographing and reformulating of the body, is designed to capture the disposition of the material body in space, that is the body considered as *res extensia*, a material displacement of mathematically specifiable space. Hence it registers not only the intelligible surface of the body as drawing and photography can do, but also its spatial organisation as a registered (that is, gridded in three dimensions – xyz) geometric volume.

Consequently, this imaging method makes not a map of the body, a surface rendition, but a model, a spatial homologue, which reproduces the inside/outside volume of the body within itself. It can demonstrate this inside/outside as a simultaneity rather than a sequence, holding together surface and depth in a single image entity. In doing so it seems to proffer a solution to the confusion between container and contained evident in the book-based atlas, by making container and contained explicitly isomorphic. The visual archive which is the VHP appears to be self-contained, a content which is coextensive with what contains it, like the archive of the human body itself is imagined in anatomy, parts which make up a (necessarily decomposable) whole. This apparent isomorphism, which in fact only defers the problem of container and contained and shifts it onwards to the 'container' of the screen and the computer itself, contributes to the status of the VHP figures as surrogates rather than maps, a status discussed below.

2 POINT OF VIEW, ANIMATION AND MORPHABILITY

This copied space is not static, like a hologram, but can rather be traversed from mobile points of view, rotated, morphed and animated in a number of ways. The VHP visually replicates bodily organisation without presenting any of its material intransigence. The interior flesh can be visually traversed in any fashion, the point of view can literally move *through* the virtual flesh at will, constituting flesh as pure spectacle, without density, recalcitrance, material consistency or self-enclosure. Hence, interior structures like the skeleton, the colon, the oesophagus, and the arteries can be navigated using 'flythrough' software, which visualises the data 'as if the viewer's eyes were freely mobile inside the … lumen' (Hong *et al.* 1995: 26). The point of view is that of a pilot in a tiny space ship, looping and zooming through the enclosed, walled spaces of the body's interior with complete mobility, able to 'land on an object in order to understand its surface structure, or to enter an object in order to understand its internal structure' (Ackerman 1991: 14). As spectacle it enables points of view which are quite unobtainable in dealing with actual bodies, either living or dead. This facility is acknowledged in the technical literature, for example in an article dealing with virtual endoscopy. Using a virtual reality system, the endoscopist can

> simultaneously visualise the anatomy and manipulate the viewing orientation in a realistic way. In fact, virtual endoscopy provides viewing control and options that are not possible with real endoscopy, such as direction and angle of view, scale of view, immediate translocation to new views, lighting and measurement. ... There are many body regions not accessible to real endoscopy that can be explored with virtual endoscopy ... including the heart, spinal canal, inner ear, bilary and pancreatic ducts and large blood vessels.
>
> (Robb 1996: 2)

If the Renaissance anatomists imagined the body as a new world to be mapped, virtual anatomists imagine it through the history of space exploration, inner space to be traversed by miniaturised bionauts, like the protagonists in *Fantastic Voyage*, and through the swooping, vertiginous optics of flight simulation. The interior space of the body is treated as empty and inertia-less space, an 'outer space' within the corporeal envelope. When the technologically assisted eye of the surgeon travels into the cavities of an actual body using an endoscope, the view conveys the labour of penetration to some extent, the vulnerable spasms and wetness of the interior flesh as it struggles to accommodate or reject penetration by the endoscope. This labour is absent from the 'flythrough' point of view, which is able to skim through the space of the body as *hollow* space, clearly demarcated from its envelope.

In addition to these mobile points of view the body image itself can be animated, providing body movements and fluid dynamics so that blood can be made to flow in the arteries, the body can bleed, the limbs can be made to flex. Attempts are being made to program in muscle torsion to make human movement modelling more plausible, and surgical simulation programs are currently in use which give the user the feeling of tissue resistance in the image. A virtual scalpel cuts into a rotatable image which registers different degrees and kinds of drag, resistance and texture to the user through a haptic feedback system. Moreover, as digital data, the body-image is fully morphable. Digital images have the quality of limitless mutability without loss of information density. They can be repeatedly changed, modified and manipulated without a loss of quality and without limits. The virtual anatomy is open to almost any kind of manipulation, because any action performed upon the body is reversible and free of consequences. As mathematical data it can be almost infinitely segmented and then reformulated. The body can be *repeatedly* dissected or penetrated, and each time resected; organs and other structures can be removed and then seamlessly replaced. Its spatiality can be endlessly deformed and then returned to an appearance which mimics the appearance of bodies in actual space, without loss of image quality. In fine, it can present the 'look' of flesh without any of the resistance and opacity of flesh, virtual flesh which acts like empty, abstract space.

3 PHOTOREALISM

The VHP bodies are 'photorealistic', in the sense that they are imaged in visible-light spectra, rather than in the range of other spectra and kinds of information used to image bodies in other forms of computer-based three-dimensional imaging. As I outlined in Chapter 1, computed tomography works through the conversion of a wide variety of electromagnetic (MRI), sonographic (ultrasound) and radiation-based (CT scan, radioisotope imaging) information into visual models of the corporeal interior. Each of these techniques anatomise the body in a particular way. The CT scan produces a detailed, colour-enhanced, cross-sectional image, good for the imaging of small bones and soft tissue, while MRI produces ghostly, aqueous images which do not register bone, useful for seeing tissue like the spinal cord within the spinal column (Sochurek 1987). Each of these imaging methods has specific clinical uses, but they also foreground the extent to which they can only be produced through extensive amounts of technical mediation. The images they produce are partial, specific, noisy, sometimes ghostly or aqueous, as with MRI, sometimes grainy, as with sonography. The viewer cannot simply see the 'thing in itself' and their use in surgery or clinical settings present multiple problems of interpretation before they can be applied to particular human bodies. The technique of the VHP is designed to overcome this necessity for interpretation, to generate a model of the body's interior which dispenses with a sense of mediated vision, yet which can still be used as a benchmark in relation to other forms of computerised vision. It produces a 'photorealistic' image, captured in the same spectrum as photography and the human eye, so that the body data appears with the colour values and resolution associated with photographic media.

These media qualities mark the VHP off from the qualities specific to the book-based atlas, and give it a quite different technical valence. While book-based anatomy can only work in the domain of sequential and static surfaces, virtual anatomy enabled the rendering of animated, structured depth. It can act not only as a *map of* actual bodies, but also as a *surrogate for* actual bodies. Two of the primary uses for the VHP are as a dissection surrogate and a surgical rehearsal surrogate (Hohne *et al.* 1996). In dissection rehearsal, anatomy students or surgeons can 'cut' into the flesh of the body with a cursor/scalpel in a fashion that mimics the spatiality of classical anatomy, treating the body as a series of layered structures which can be revealed as surfaces, or discrete organs which can be removed from the body cavity and dissected. (See Figure 3.3.)

Their photorealistic, volumetric quality means that such incisions produce views of interior structures which resemble those produced by a similar procedure on an actual corpse. (See Figure 3.4.)

Surgical rehearsal or simulation involves practising certain techniques, the use of probes and catheters, endoscopes and the like, on the virtual body, as preparation for working on a patient. Users can, for example, rehearse a craniotomy by cutting arbitrarily shaped holes and removing tissue layer by layer from the virtual skull, plotting an access path to be followed by a probe (Hohne *et al.* 1996: 29). Paramedical staff can receive training for certain procedures, the administration of injections for example, using a virtual model based on the VHP data. These surgical applications are discussed in much greater detail in the following chapter.

TOMOGRAPHIC SPACE, DIGITAL INSCRIPTION

These medium-specific qualities make the VHP a highly productive technology for medicine, able to be used in multiple sites and multiple

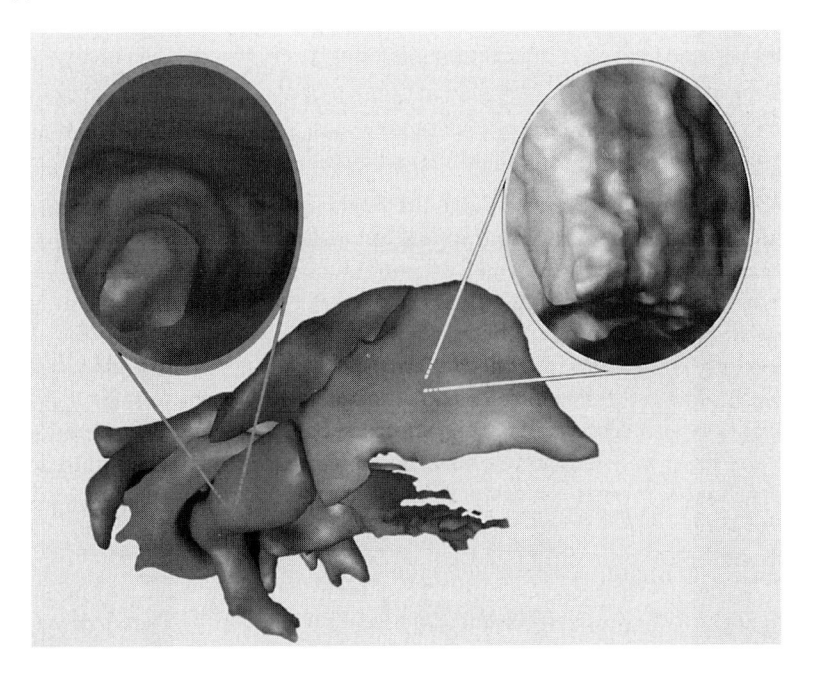

Figure 3.3 Heart model computed from Visible Male data (image courtesy of R. Robb, Mayo Foundation)

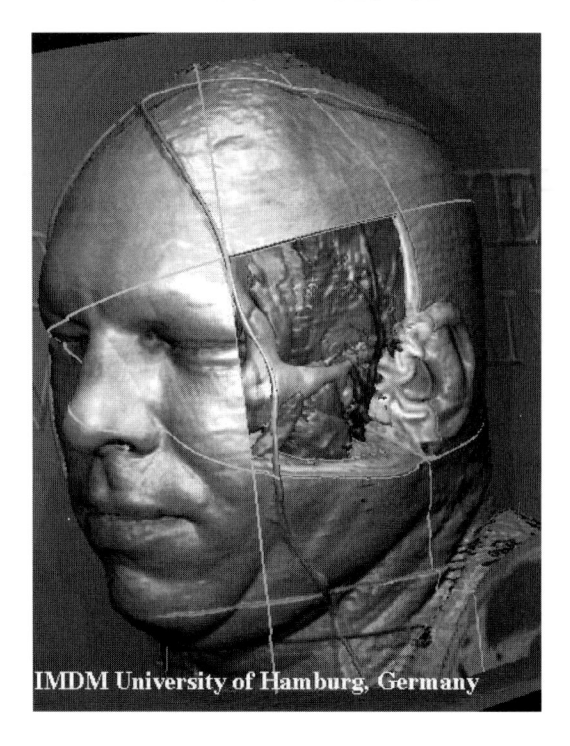

Figure 3.4 Visible Male head showing interior structures (image courtesy of Institute for Mathematics and Computer Science in Medicine, University Hospital Eppendorf, Hamburg)

modalities to develop new forms of medical imaging (the flythrough), to better manage time and planning within surgical procedures (rehearsal), and to carry out modelled research, testing new procedures on interactive, volumetric data. The VHP data-body can perform in these spectacular ways because of the particular mode of sacrifice involved in its production. Instead of a corpse dis-integrated according to the demands of the book's space, the bodies used for the VHP were sacrificed according to the demands of digital inscription and tomographic space. The VHP is a visual text produced by literally reworking the body's materiality according to the logics of computer storage and computer vision, and so rendering it as a system of traces workable and readable through the medium of computed space.

Here the cryosectioning procedure can be recast as itself a kind of reading and writing practice, a literacy whose logic is determined by the material organisation of computer vision and visualisation. The cryosectioning freezes and then effectively annihilates the corpse in order to capture its volume and dense structure as digital visual data, as code trace which composes a digital archive. The cryosectioning procedure makes the body's interior readable by converting its volume and density into a series of shallow, archivable planes. While book-based anatomy requires the body to be anatomised as an accretion of laminar surfaces, computer based anatomy demands that it be reduced to a series of flat cross-sections which can be both serially visualised and 'stored' as discreet digital files. This archival method is a literal enaction of the tomographic method of imaging; the VHP *literally* cuts the figures used in its image production in the same spatial configuration as conventional tomographic methods – CT scans and MRI. Conventional tomography *optically* 'cuts' patient bodies as fine cross-sections, through the transmission, reflection or emission of various forms of radiation. (See Figure 3.5.)

CT scans, the most widely used tomographic method, use x-ray radiation, but rather than providing a 'shot' from one angle, as conventional radiography does, the scan uses a thin beam which rotates around the entire body and takes an image 'slice' through a cross-section. The images produced by CT scans are a series of such slices which move gradually through the body's length. Hence tomographic imaging produces a visual archive which records not just the appearance of the body but also its volume, through the computer's ability to retack the succession of optical slices with registered precision, maintaining relations of geometric extension and scale within the image itself.[12]

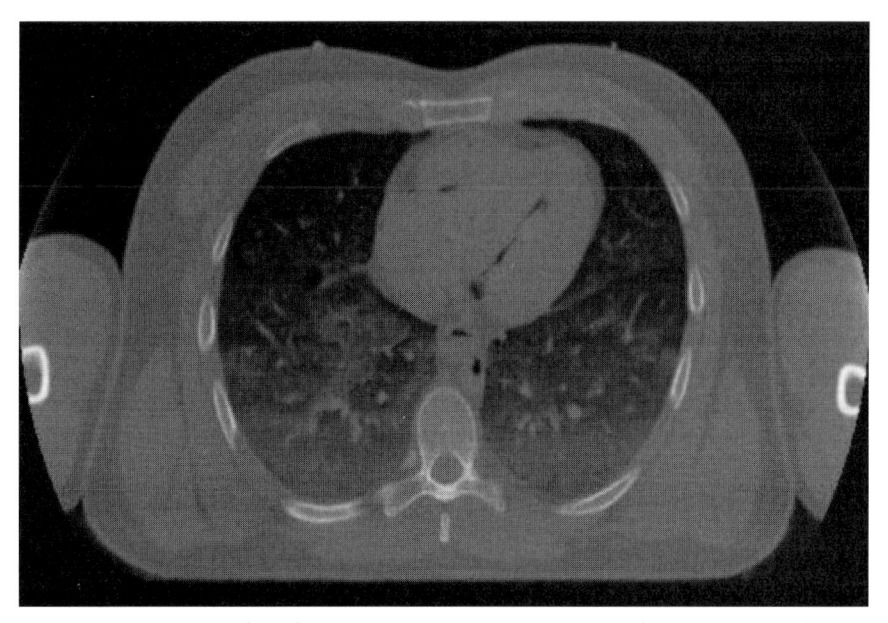

Figure 3.5 CT scan of Visible Male thorax (image courtesy of the National Library of Medicine, USA)

The VHP effectively substitutes a material incision for the radiation 'incision' used by the CT scan, a substitution which enables the visual data surrendered by the body to be captured in photographic spectra, although not as analogue light-sensitive traces in emulsion but as digital data. All information manipulated by computer must be rendered as digital data, as binary code which mathematicises all forms of phenomena – numbers, text, images, sound, radiation – as strings of bits, combinations of zeros and ones. By moving from the continuous, analogue realm of the book into the discreet, digital realm of the computer, the VHP produces a visual text of the body which is also a structure of mathematical data, digital code.[13] Hence the VHP is, in effect, a mathematical description of corporeal space, a point-to-point mapping which substitutes the navigable space of data for the dense intractability and opacity of flesh. It is this point-to-point methodology which drives the appeal of the VHP as surrogate object for so many domains of medicine and biotechnology. As trace it is understood to be exhaustive, to leave behind no significant residue, but rather to capture all that is essential for a medical vision of the body, and to port it as positive content into a new medium, the medium of data.

Nevertheless, the VHP's appeal as surrogate resides in a paradoxical tension between the process of sacrifice and the form of the trace so produced. The attractiveness of the VHP trace for medicine resides, as I have already described, in its qualities of immutable mutability, traversability, scalability, volumetric complexity, photorealism and tomographic registration. All of these qualities fit it for use as simulation of a living body, a volumetric image surrogate and virtual test subject upon which procedures can be trialled and practised. At the same time this surrogate is produced not through a doubling of presence, a copying, but through the rationally ordered destruction of both being and matter. The condition of possibility of the VHP data-bodies is precisely the making-absent of the bodies that it must use up in the process of its creation. The VHP is, at bottom, a mathematical description of space which does not record a body's disposition in space but rather the process of its abolition. The sequential destruction of the actual flesh makes the virtual flesh traversable, mutable and reproducible, because its abolition is the sacrifice necessary for its transubstantiation into visual data and mathematical code.

REPAIRING DEBTS

The process of production of the VHP, and its range of uses (discussed in more detail in the following chapter) are productive for biomedicine within two separate, but related, economies. One of these is a biopolitical economy. As I have demonstrated, all anatomical production is based on sacrifice, the death and dis-integration of some in order to enhance the health of others. In the recent history of anatomy the earlier, steep hierarchies of biopolitical value between the anatomised and those who benefit from anatomy have been attenuated to some extent. That is, in the USA, Britain, Australia and elsewhere in the industrialised democracies, the bioethically supervised procedures for anonymous donation of bodies to science have largely replaced the forced appropriation of the corpses of the poor and mad, of slaves and other marginalised groups, for anatomisation. The donation of the anonymous Visible Woman is an instance of such procedures. Nevertheless, the sacrifice of Jernigan's corpse suggests that this older biopolitical economy has not been completely effaced. Practices in other areas of medicine, notably a series of recent cell line controversies,[14] or the use of third-world populations to test AIDS vaccines, point towards the continued operation of an economy of knowledge which utilises the bodies of the less valuable as test technologies to assist the more valuable.

The uses being made of the VHP are also symptomatic of such an economy. While the VHP is itself in the public domain, and the National Library of Medicine has endeavored to make it as accessible as possible to a wide range of research groups, many of its users are focused on the development of new markets for privatised medicine, particularly within the highly privatised and profitable US medical system. As Howell's (1995) careful historical account of medical imaging in US hospitals clearly demonstrates, high-end medical imaging secures investment and research funds, and is institutionalised within hospitals, only if it can be used to create new markets and secure high fees from patients able to pay for advanced kinds of diagnosis and surgery. Only those hospitals able to invest in high-end clinical, anatomical and surgical viewing stations and transmission networks, and in highly trained technicians and radiographers, will be able to deploy this form of imaging in the foreseeable future. To this extent the bodies of Joseph Jernigan, executed criminal, and an anonymous Maryland woman, have provided the raw matter which will enable the preservation of the lives of those who can benefit from high technology medicine – the heavily insured in the increasingly privatised medical systems of the industrialised democracies.

The second economy of the VHP's productivity could be described as biosemiotic. The VHP data-bodies are seductive for medicine because, as I have already observed, they are morphable, reversible, reproducible and indestructible. These qualities have extensive practical value for medical imaging and modelling, but they also work to repair in representation the disavowed debt which biomedicine owes to the bodies used in imaging. While the bodies used to produce the project may have been literally annihilated, their imagos enact the process of repair again and again, and demonstrate their own indestructibility. They are exemplary test subjects, able to be used and reused, cut up and seamlessly repaired, fragmented in an infinite series of ways and reconstituted. Like all digital objects they can sustain transformation without loss, change without cost, nothing done to them has consequence because everything can be returned to initial conditions.

Hence the VHP imagos are appealing for medicine, not only because they seem to erase medicine's indebtedness to the bodies used in the VHP, but also, perhaps, because they erase in representation a debt which all living bodies owe to time and death. After all, the process of production of the VHP, its annihilation of two bodies, bears witness to the indelibleness of flesh, its non-reversibility. Once it is marked, its marks cannot be erased. All pain, all experience, all illness and all violence leave a trace in flesh, changing it forever. In this sense flesh is historical, conditioned by the passing of time which eventually brings its death and dissolution. If flesh can be said to be a medium, the traces which traverse it always inscribe a certain loss to history, marking the temporal limits of the body and the unrecoverable debt it owes to its own eventual disappearance. When flesh is transubstatiated into data, the negentropic world of the computer seems to suspend this debt. The Visible Human Project is seductive for medicine because it presents an imaginary anatomy in which the debt to organic time has been cancelled.[15]

4

VIRTUAL SURGERY
Morphing and morphology

The Visible Human Project constitutes a decisive move within medical knowledge away from the bibliometrics of the anatomical text and toward the cybermetrics of the anatomical simulation. The performative flexibility of the figures, described to some extent in the previous chapter and discussed in greater detail here, is an aesthetic yield produced by the performative flexibility of visual data itself. This economy of simulation produces, in turn, multiple use values in the field of diagnostic and surgical imaging. The two data-figures are currently being adapted and utilised in a rapidly expanding range of clinical and pedagogical applications, circulating as new, universal anatomical norms throughout the ever expanding networks of cybermedicine. This move is not merely a matter of better or different modes of representation. Rather, it has important implications for ways in which medicine can work the relationship between the body's endosoma and exosoma, surface and depth. In the previous chapter I argued that the anatomical text is a technology which writes the body *as a technological ensemble in workable space*, an assemblage of organ-tools whose order can be both analysed and adapted to prosthetic reordering. The anatomy text opens the body out as a system of instruments, with potential relationships to other forms of instrumentation. The transformations of anatomical practice involved in the Visible Human Project generate new forms of access and prosthesis, new ways to plot trajectories through the complex partitionings and flows that constitute the bodily interior. It alters the ways in which biomedicine can address the body and engage in therapeutic and normative interventions in the macroanatomical morphology. It refines and potentiates ways that the space of living bodies can be worked in relation to the space of visual computation. This chapter examines this process of transformation, with a particular focus on the relationship between tomographic imaging and surgical practice.

Everywhere in the world of medical imaging research the VHP data is being used to extend or refine the already extensive repertoire of computed medical imaging, and as a surrogate object which is transforming earlier forms of vision. At the very threshold of induction into medical ways of seeing, medical schools throughout the industrialised nations are using the Visible Human data in their anatomy curricula.[1] In the first instance the VHP provides the novice anatomist with a virtual rather than textual map of the cadaver being dissected, a map which presents itself in three rather than two dimensions and which can be used to stand in for the cadaver in various ways. He or she might rehearse dissection techniques on the virtual body before carrying them out on the corpse, and refer back and forth between virtual and actual corpses, in order to learn (that is, order) the actual body anatomically. A more radical proposal involves the *replacement* of the cadaver on the dissection slab by the virtual cadaver on the workstation screen, so that experience of dissection is only ever virtual, and an actual fleshly cadaver is never encountered. The possibility that the VHP data might act as a substitute for actual cadavers rather than an aide in their study has caused some anxiety in medical circles. Csordas reports that, at the 1996 'Medicine and Virtual Reality' conference,

> The primary debate [was] over whether these methods are to be used to complement or to replace conventional dissection in anatomy education. Traditionalists are resistant to the idea that medical students would not have the hands-on experience of work with real bodies in what is implicitly sacrosanct as a rite of passage in medical training. Innovators point out that actual cadavers are increasingly short in supply in comparison to the infinitely reusable [VHP data], and that in any case most physicians other than surgeons will never have occasion to work on the insides of their patients.
>
> (Csordas 1998)

It is evident here that the VHP's capacity for surrogacy, to act 'as if' it is a body, presents a certain challenge to the centrality of the 'actual' cadaver in medical education, and by extension in the medical imagination more generally. The VHP seems to further displace the centrality and immediacy of the therapeutic touch, the hands-on experience which is here equated with an unmediated encounter between hand and flesh. The medical profession's anxiety about this arises from a general cultural anxiety regarding the capacity of computer-mediated communication to defer and displace face-to-face relations, the challenge it poses to a nostalgic investment in authentic presence. It also registers the dynamic tension between the humanism of medicine's rhetoric and the technically driven nature of its practices.

Another major area of application for the VHP is in virtual diagnosis and surgery. This term encompasses a diverse range of practices which use the volumetric properties of computer vision to develop more accurate methods of diagnosing pathology and carrying out therapeutic surgical interventions. As I have already described, CT, MRI and other kinds of scans can be carried out on patients and the data reconstructed in the same way that the VHP

VIRTUAL SURGERY

cryosections are reconstructed, making three-dimensional models of the patient's skeletal structure or internal organs. (See Figure 4.1.)

Such models can be used diagnostically, to identify pathology or deformation in the organ. They provide a volumetric, spatially accurate view, rather than the flat, axial view provided by conventional CT, MRI and other kinds of scan data. The Visible Human database is, for example, being used in the development of virtual endoscopy at a number of sites, notably in the medical faculty at State University of New York (SUNY), Stony Brook and at the Mayo clinic. Conventional endoscopy uses fiberoptic cable and tiny video cameras to visualise the interior space of certain lumens – the oesophagus, the colon, the fallopian tubes and other tubular organs. The endoscope is inserted into lumens through the mouth, anus or navel and projects a video image of their interior surfaces on a closed-circuit television screen. Conventional endoscopy is used both diagnostically, for the identification and inspection of pathological structures (polyps for example), and as part of some surgical procedures, like laparoscopy, where keyhole surgery is performed on the internal organs of the abdomen. By contrast *virtual* endoscopy provides a means of imaging the interiors of lumens without direct penetration of the body. In the words of Richard Robb, one of its primary proponents, virtual endoscopy

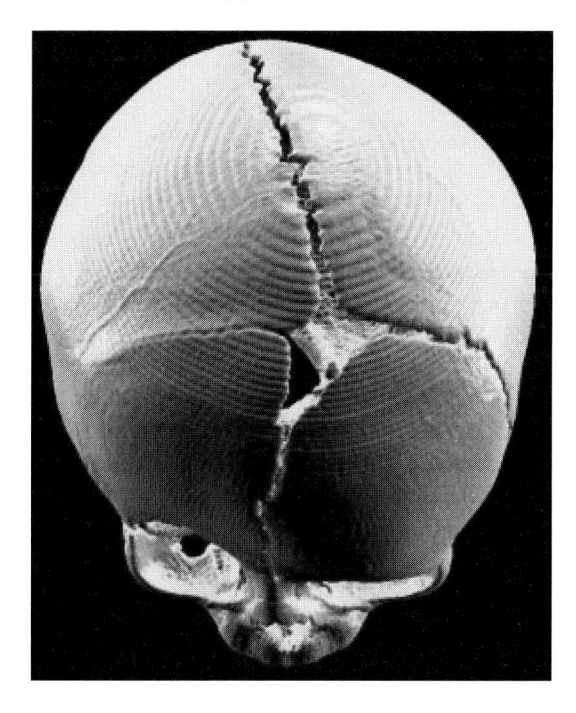

Figure 4.1 CT reconstruction showing infant skull (image courtesy of M. Ross, NASA Ames Research Centre)

is a new method of diagnosis using computer processing of 3-D image datasets (such as CT or MRI scans) to provide simulated visualisations of patient specific organs similar or equivalent to those produced by standard endoscopic procedures. Conventional CT and MRI scans produce cross section 'slices' of the body that are viewed sequentially by radiologists who must imagine or extrapolate from these views what the actual three-dimensional anatomy should be. By using sophisticated algorithms and high performance computing, these cross-sections may be rendered as direct 3-D representations of human anatomy.

(Robb 1996: 1)

Here the VHP data is being used to 'develop, test, and compare 3-D visualization and image processing methods, and to evaluate the effectiveness of these methods, including endoscopic simulation, for eventual applications in clinical diagnosis and therapy' (Robb 1996: 2). The VHP provides images in the same formats (CT and MRI) as those used for gathering patient data, and also makes a photorealistic, co-registered image available, which can be compared with conventional endoscopic, photorealist images. Hence it provides medical imaging researchers at multiple sites with an already prepared and agreed upon standard dataset, which can be used to benchmark new virtual endoscopic methods against already standardised endoscopic methods. Techniques for data gathering, volume visualisation, surface rendering and navigation can be tested out on the VHP data and the different imaging modalities compared with each other, with standard endoscopy views and with the overall body orientation. The standardisation of imaging techniques is a necessary part of their becoming widely used as reliable diagnostic and surgical planning practices (Cartwright 1995). This is one of the primary functions performed by the VHP data in imaging research, a means to 'refine and validate simulations for routine clinical use' (Robb 1996: 2), to provide a '"ground truth" measure for validation ... against which image processing effects using the CT and MRI data can be calibrated and judged' (Robb 1996: 5).

The VHP data are also being used to generate surgical simulation techniques, providing a surrogate body part on which surgeons can be trained, using haptic feedback systems. The novice surgeon learns certain kinds of procedures by sitting at a workstation, wearing data goggles and holding surgical instruments attached to force feedback devices that provide the feel of interactive resistance to the surgeon's hand when the virtual instrument encounters the virtual object. The surgical instruments, like the mouse in relation to virtual text, track the action and position of the hand and reproduce it in virtual space. The virtual representations of the instruments can touch, cut, grasp and suture the virtual body part, which has different degrees of density, pliability and rigidity programmed into appropriate regions to provide convincing 'feel' (Playter and Raibert 1997). At time of

VIRTUAL SURGERY

writing the VHP data are being used along these lines for the training of rheumatologists using a virtual knee (see Figure 4.2),[2] for simulation of craniofacial surgery,[3] and vascular catheterisation techniques.[4]

The data provides a volume upon which to practice the proprioceptive skills of surgery as they are deflected through virtual instruments and screens, and as an exemplar of normal anatomy. The catheterisation program, for example, allows physicians to practise the insertion of flexible tubes through the blood vessels of a patient's body, a procedure which is guided by a real-time fluoroscopic display mapping the vasculature and showing the location of the catheter head. The Visible Human simulation tracks the trajectory of a virtual catheter through the vasculature of the Visible Human data, providing real-time visual and force feedback to the user. In the case of craniofacial surgery the VH data are used to provide a surrogate skull, a deformable model of the head which can be used to project the effects of various surgical interventions.

> The important role played by the Visible Human dataset is to provide us with normal examples of craniofacial features, both skeletal and soft tissue,

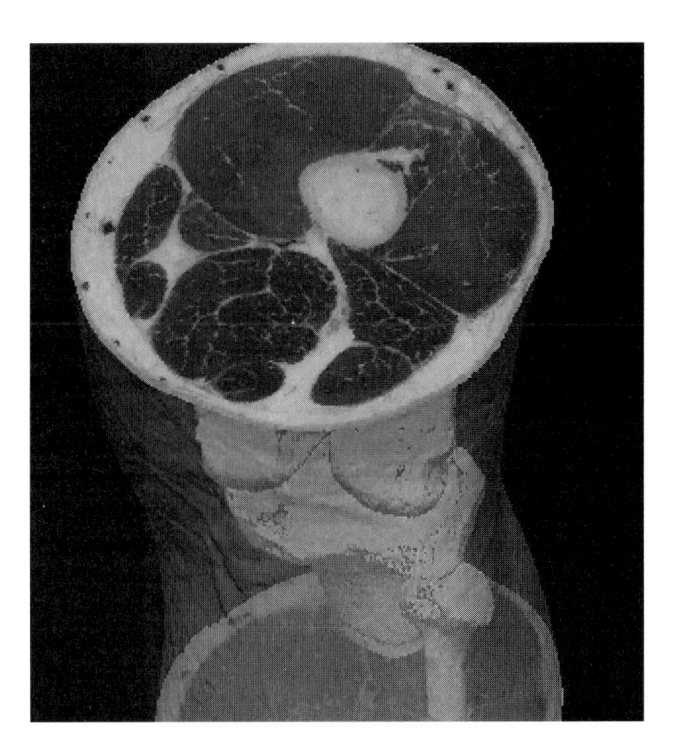

Figure 4.2 Three-dimensional reconstruction of knee using VHP data (image courtesy of William Katz, Varian Medical Systems)

for comparison with patients suffering craniofacial trauma or disfigurement. This will provide the novice with information concerning the degree of change involved in the affected patients.

(Ross 1996: 1)

The novice surgeon can practice craniofacial procedures on the dataset, using virtual drills and scalpels, carrying out simulated bone grafts, bone reconstruction and soft-tissue work. The head and face can be re-imaged based on this surgery, projecting the ways in which the skull and face will heal in the wake of the surgical intervention. 'If the outcome of the virtual surgery is not desirable to the surgeon, the virtual process can be repeated until satisfaction is obtained, without touching the patient' (Ross 1996: 1). The data can be used to model and project subsequent growth and post-operative deformation in an infant's skull, anticipating undesirable outcomes of surgical interventions into still-growing tissue. Surgical simulation is advocated for surgical education as a way to

> train sensory-motor skills, perceptual tasks and cognitive decision making, all in the context of performing relevant surgical procedures. In addition to allowing medical students to practice routine procedures, surgical simulators will allow students to encounter patients with rare medical conditions or unexpected complications and practice techniques to handle these situations. Using today's opportunistic methods, only very experienced surgeons are exposed to the wide range of conditions and complications that can occur. ... Surgical simulators could [also] be used to assess surgical performance and play a role in certification.
>
> (Playter and Raibert 1997: 117)

VHP data are also being used for surgical planning and rehearsal, techniques closely related to surgical simulation. Surgical planning refers to the plotting out of incisions and techniques designed to minimise tissue damage and maximise the 'targeting' of the area of interest. Here, one use for the VHP data is to serve as a model of the slippage between different imaging modalities. For complex operations images may be gathered in ultrasound, CT and MRI formats and used in concert with each other. However, given the different capacities of these three methods to image different kinds of tissue, and the probability of patient movement and tissue deformation between and during scans, the data-sets are often difficult to co-register with each other, and to the patient in the operating theatre. The Visible Human data facilitate comparisons across these different modalities, because the anatomic (photographically based) data allow easy differentiation between tissue types (bone, tendon, muscle, fat, etc.) and registration with the Visible Human MRI and CT data. Hence the Visible Human data can be used to measure relative transformations between different modalities, and generate a set of test data and matching algorithms that can be utilised to assist interpretation between different kinds of patient data (Edwards *et al.* 1996).

Surgical rehearsal uses simulation techniques to assist in surgical planning. It requires the gathering and rendering of patient data into a three-dimensional virtual model, which is then used to simulate the surgical path into the patient's body. The VHP data are being used in this context to provide an anatomical 'framework' for specific data about patients' organs, and as supplementation for the deficiencies of other imaging methods, a use discussed later in this chapter. In one example of this procedure the Visible Human Male pelvic girdle is used as a skeletal framework for a model of the bladder, prostate and adjoining organs, including a tumour, computed from a patient MRI scan, and then registered and inserted into the VHM pelvis. (See Figure 4.3.)

> The VHM pelvic girdle model serves as a framework for studying the patient-specific models of the prostate [because] pelvic bones are not available on MRI scans. The urological surgeon can use such displays in a real-time virtual environment to critically study the anatomic relations preoperatively.
>
> (Robb 1996: 7)

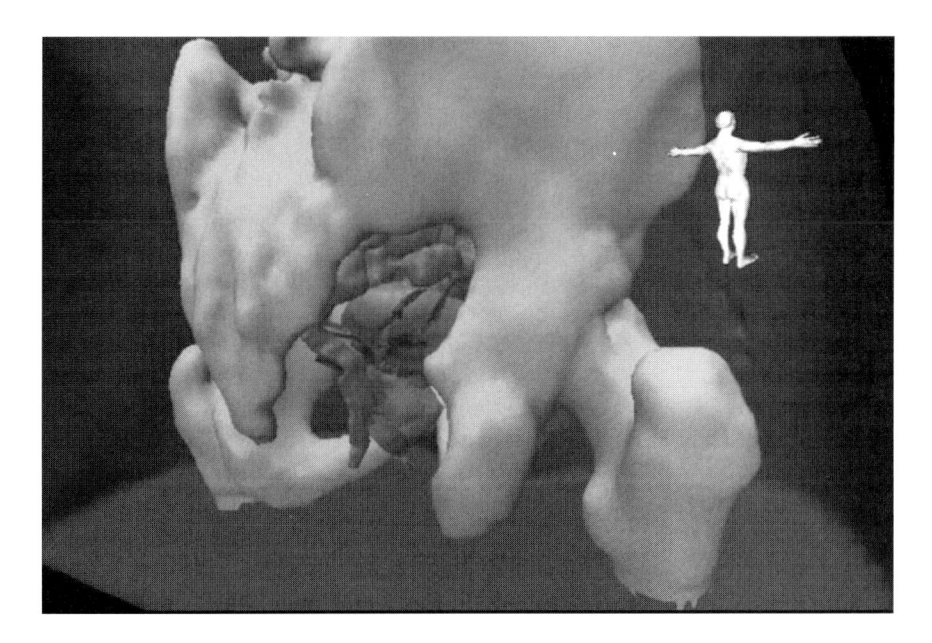

Figure 4.3 Visible Male pelvis with prostate gland from cancer patient (image courtesy of R. Robb, Mayo Foundation)

The wide range of uses for the Visible Human data, a range in constant extension, resists neat summary. Some of the uses I have already described are concerned with the enhancement of the surgeon's haptic skills, as these are increasingly inflected through cybernetic systems of visualisation and feedback. Some are concerned with the role of time in surgical practice – the projection and rehearsal of procedures, the modelling of instrumental cause and morphological effect. In this chapter, however, I want to focus on one category of uses for the VHP – its ability to manage, refine and normalise imaging technologies dedicated to televising the endosoma, the projection of the body's internal morphology as externalised image. Such projection is nothing new in the history of medicine. As I will demonstrate, it is coterminous with the practice of (successful) internal surgery and a crucial supplementation to anatomical imaging in surgical practice. The Visible Human Project is continuous with a history of such projection. At the same time it is strongly implicated in current transformations in surgical practice, transformations which precede the VHP but which its advent will further potentiate and systematise. I am referring here to various attempts to move surgery away from 'open surgery', the laying open of the bodily interior through large incisions, and towards minimally invasive practices like keyhole surgery, and non-invasive therapeutics like high-intensity focused ultrasound and laser. That is, new surgical techniques attempt to work the interior of the body telematically, rather than through a direct visual and haptic encounter of the surgeon with the exposed corporeal interior. Instead of bringing the patient's interior to the light of the surgical theatre, such surgery relies on the virtual light of the scan, and the intelligibility of tomographic space to make diagnoses and undertake therapeutic interventions. The VH data are being used to research and develop techniques which defer, avoid, or reduce the frequency and duration of open surgery, by facilitating a more precise, homogeneous mapping of interior space as volume, prior to surgical incision.

In what follows I want first to demonstrate the dependence of all internal surgery on telematic practices, by sketching out a brief history of the projection of living endosoma, its exteriorisation as image as a crucial move in its instrumentation. Current tomographic methods and the Visible Human Project are only the most recent innovations in this history of telesurgery, despite attempts by practitioners to claim telesurgery as a new development, only beginning with the application of computers to medicine.[5] Telesurgery in the latter's terms describes kinds of surgery, now in prototype, in which the surgeon's and patient's bodies never meet in proximate space. Instead 'patient data', including vital status monitors (heartbeat, blood pressure), and visual data collected using real-time endoscope cameras and tomographic data (CT, MRI, PET) is fed to a screen interface, a computer or video, which may be in the next room or 2,000 miles away. The surgeon operates remotely. Computer-controlled instruments work on the patient.

These instruments are manipulated by dummy handles with force feedback to give the surgeon 'feel'.

Telesurgery is thus an instance where there is no tangible encounter between patient and surgeon. There is an absence of direct unmediated touch or face-to-face meeting which, as I remarked earlier, has provoked a certain anxiety among a medical community for whom unmediated touch is emblematic of a certain humanist ethics. For example Howell's (1995) history of the introduction of scientific technology into US hospitals at the turn of the twentieth century writes the history of the x-ray as a retreat from the therapeutic and diagnostic power of the physician's touch, the x-ray as an interruption and barrier between patient and physician. This anxiety shows signs of nostalgia however, a longing for an idealised medical past replete with an unmediated therapeutic intimacy between doctor and patient, an absence of instrumentation which is equated with a fullness of communication. Telesurgery is celebrated by its proponents as a completely new practice, but it is, rather, the most recent restaging of a theatre of surgery in which projections, screens, displacements of location, and the mediatisation of touch and vision have always been involved in the production and management of the corporeal interior, and indeed make internal surgery possible at all. While palpation, the manual probing of the patient's body, is an important procedure for the location of organs or trauma, and in itself places limits on the possibilities of remote surgery (Bowersox *et al.* 1998) this surgical touch cannot proceed without the supplementation of haptic and visual technologies. The touch of the surgeon is never uncontaminated by the complicating and mediatising dis-locations of instrumentation. As I will demonstrate, all viable surgical intervention evokes and relies upon some kind of telematic space, and much progress in surgical techniques has depended on the application of new forms of teletechnology, new modes of projection and encounters not between surgeons and patients but rather between surgeons and screens. Then I want to consider the particular cybermetric and tomographic qualities of the Visible Human data, and the ways in which it lends the surgical patient's body to new orders of instrumentation and translates it into new kinds of technological ensembles.

EXSCRIPTION: TELEVISING INTERIOR MORPHOLOGIES

Internal surgery, the instrumental address of the living body's internal organs, only became possible at the point when medicine developed *exscriptive* imaging technologies. I am using Lippit's (1996) term here to refer to technologies that write the living body's interior morphology as externalised image-objects, traces or reflections projected onto screens. Exscriptive technologies are techniques for imaging the living body, a visual 'dissection' which does not involve an actual incision. This is what distinguishes diagnostic imaging from the anatomical techniques discussed in the previous chapter. Anatomical imaging involves a literal dissection, cutting into the flesh in order to produce it as a system of illuminated surfaces, techniques which are necessarily techniques of the corpse. The therapeutic vivisection of the living body, internal surgical intervention designed to preserve life, necessarily depends upon exscriptive technologies like the endoscope, the x-ray and the tomograph, which can project the body's interior as exterior image.

Internal surgery did not become viable as a therapeutic practice until the mid-nineteenth century, when new techniques developed for the increasing visual instrumentalisation of the medicalised body. Until then, the discipline of operative surgery confined itself to a very limited range and small number of operations – the draining of abscesses, amputation of limbs, excising of external tumours and suturing of fistulas were the most common. Open operation, the cutting open of the body's cavities, was rarely attempted (Cartwright 1967). While surgeons may have had a knowledge of dissection anatomy, this in itself did not offer a direct guide to the location of disease within the opaque volume of the particular, living body. As normative models, anatomies can only act as general maps, giving no exact indications regarding the relations of surface to depth found in any particular, living body. In the absence of any means of determining the specific location of a bullet or tumour for example, any attempt to cut into the body risks dangerous and unproductive wounds.

Developments in the techniques of illumination in the mid-nineteenth century allowed Victorian physicians to develop instruments that could introduce light into the body's structures and cavities, and project illuminated images outwards: ophthalmoscopes, which refracted a light source into the eye to illuminate its interior; laryngoscopes, which allowed viewing of a mirror image of the larynx and trachea through a complex system of refracted light and mirrors; early endoscopes, which introduced carbon-filament lamps into lumens – the stomach and sinuses most commonly – 'a light brilliant enough to outline their shape and contents through dense covering tissues' (Reiser 1978: 56). The developments of these various scopes produced ways to telegraph the body's internal structures, rendering them externally, remote from their internal site, as mirror images (the laryngoscope) or shadow images (the early endoscopes and ophthalmoscopes) which used the body's exterior as its own screen. Each exploited the body's configuration as itself a part of an instrumental ensemble, a system of enclosures which could be opened out and visually projected, and each yielded rapid diagnostic and surgical results. The laryngoscope permitted surgical treatment of tumours on the vocal cords for example, providing a means for physicians to manoeuver surgical instruments into the throat with precision. Like the anatomical dissection of the corpse these techniques allowed the body's depths to be externalised, effecting a reconfiguration of the relations of surface to depth. As Reiser puts it, with the invention of the laryngoscope, 'the voice box ceased to be an internal organ to physicians' (Reiser 1978: 53). Instead it became available for direct diagnosis and surgical intervention, in a fashion previously restricted to pathology on the body's surface.

So the introduction of technologies that altered the form and extent of visualisation rapidly provoked new methods for touching, cutting, cauterising and suturing, methods addressed to the newly externalised organs and sections of the body. As Keller (1996) points out, new techniques for the projection of what she terms the 'biological gaze', techniques like the microscope for example, always invite the creation of new techniques for working within the new visual field, new haptic technologies that reconfigure and extend the experimental and therapeutic touch. All new scientific forms of imaging involve a reorientation of proprioceptive relations between eye and hand, a retooling of their interdependence so that they co-ordinate at a different scale, distance or depth.

RADIOGRAPHIC EXSCRIPTION

The advent of the x-ray, the technology that effectively established the logic of most subsequent twentieth-century imaging technologies, provoked a similar transformation in surgical practice.[6] X-ray radiation passes through flesh but is blocked by bone and other dense structures. Directed at a body part, it produces an image which scarcely registers flesh, and shows bone as a darkened mass on a photographic plate or fluorescent screen. X-rays produce an image which superimposes the outline of the body's surface on the opaque shadows of its interior in an indeterminate series of receding graphic depths. As photographs they are a 'simultaneous view of the inside and the outside', a collapse of surface into depth which produces 'at once images of a three-dimensional flatness' (Lippit 1996: 34–5). The surface of the body, its demarcation from the world, is dissolved and lost in the image, leaving only the faintest trace, while the relation between depth and surface is reversed. Skeletal structures, conventionally thought of as located at the most recessive depth of the body (Schiebinger 1987), appear in co-registration with the body's surface in the x-ray image. Hence skeletal structures are externalised in a double sense: the distinction between inside and outside is suspended in the image, and the trace of the interior is manifest in the exteriority of the radiograph, the artefact itself. As Lippit describes it,

> The ability to simultaneously expose the inside and outside of a thing, to retain the object's surface even while probing its depths describes a scientific phantasy as well as a scientific imperative. Unlike the microscopes and telescopes which precede it, the x-ray retains the dimension and shape of the object while rendering its inside. The imaginary x-ray document thus generates an impossible perspective. Figure and fact, an object's exterior and interior dimensions, are superimposed in the x-ray, simultaneously

> invoking and complicating the metaphysics of topology in which the exterior signifies deceptive surfaces and appearances, while the interior situates truths and essences.
>
> (Lippit 1996: 39)

If the modest medical scopes described earlier reconfigured the opacity and depth of the body in important ways, the x-ray utterly transformed them. While the scopes selectively introduced illumination into certain bodily depths and projected them outward, the x-ray introduced a form of light which no longer glanced off an inner surface to make it accessible for medical vision but rather cut through the very distinction between inner and outer. Its spectral images rendered the body's interior as irradiable *space* and illuminated scene.[7]

The light of the x-ray does not simply penetrate its object, it also projects it, moving through it until its force is interrupted by a screen. Hence the trace of anatomic structure can be both externalised and fixed as radiographs. Unlike the early scope technologies, which could only select a portion of relatively accessible anatomy, typically structures like the vagina or oesophagus that establish points of communication between the inside and outside of the body, the x-ray can externalise dense anatomical structures, without concern for their location within the depth of the body. Hence the x-ray provides a visual object which can stand in for the body in new and highly productive ways. While the ophthalmoscope and laryngoscope only admit one viewer at a time, the radiograph is iterable, and hence communal. It takes a reproducible diagnostic image which can stand in for the body of the patient, and form part of an archive. It can be circulated, studied and discussed by a number of physicians simultaneously. Comparisons can be made across different cases, and between radiographs of the same patient before and after surgery. Moreover, radiographs can be systematically compared with the existing repertoire of anatomical images, and forms of standardisation arrived at.

The advent of the x-ray very rapidly produced a yield for surgery, giving it access further into the bodily interior.[8] In concert with the practices of systematic asepsis[9] and the development of technologies which produced real-time data about patients' physiological conditions (the electrocardiograph for example), radiography enabled the dramatic expansion of open operation, surgical procedures carried out on internal organs – appendicectomy, removal of gall stones, hernia repair, thyroidectomy, the surgical treatment of stomach ulcers, and so on. Abdominal surgery, previously carried out in exceptional circumstances, became relatively common. As Starr (1982: 156–7) reports, for example, the Mayo brothers, who were among the first surgeons to install x-ray machines in their clinic, performed 54 abdominal operations between 1889 and 1892, 612 in 1900 and 2,157 in 1905.

Radiography created the conditions for these surgical incursions into bodily depth because it provided a way to project those fragile depths as a durable surface, the radiograph itself, making its interior contours available as external objects. In doing so it removed the limit previously presented by the opaque volume of flesh to precise, therapeutic vivisection. The radiograph acts as map, a device for the anticipation of, and navigation within, previously unencountered territory. Open surgery involves visually and instrumentally projecting a precise trajectory from the initial incision on the surface through flesh and viscera to the organ or region of interest, a trajectory which both requires prior visualisation and is itself a means of visualisation. Hirschauer, in his surgical ethnography, describes this process in the following terms:

> After the [first] incision, a step-by-step occupation of the patient-body takes place. The surgeon-body extends itself into the flesh: with fingers, clamps, suction tubes and cutting instruments. One layer after the other is removed, camp is pitched, and the expedition continues. The visual reduction by [draping] reveals itself as one step in a process of targeting, which is now realised by instrumental means. The operator surrounds an organ, trying to spare the neighboring parts of the body. Layers of skin and tissue obstructing the view are cut through and spread apart. Operating becomes a sequence of looking and cutting, of manipulations providing visibility for further manipulations.
>
> (Hirschauer 1991: 299)

Within itself the radiograph anticipates the surgical production of the interior *as surface*, the bringing-to-the-surface which is the act of open surgery. The surgical theatre is a space in which the usual self-enclosure of the body can be interrupted and its boundaries reconfigured, so that its interior becomes a workable surface. The complex technical panoply of the surgical theatre – the life-support machinery, monitoring equipment, detailed aseptic procedures, the viewing station for the display of x-rays or other imaging modalities – converges around the patient's body in order that its interior space can be safely exteriorised (Young 1997). Moreover the spatiality of open surgery is the laminar spatiality of the atlas, the production of successive depth and contiguous volume as serial, illuminated topographic surfaces. During surgery

> The imaginary anatomical body, which in the beginning is not visible at the patient-body, is looked for and constructed at the patient-body in the course of the constant attempt to produce its visibility. *One 'leafs through' the three-dimensional patient-body to find the two-dimensional structure of anatomical pictures.* Section after section, the proper anatomy of the ideal body is engraved on these layers.
>
> (Hirschauer 1991: 312, emphasis mine)

TOMOGRAPHIC EXSCRIPTION

These modest scopes and the more dramatic x-ray are significant for a consideration of the VHP because they lay out the early technologic of exscription and the ways it enables the instrumental address of the bodily interior. Each of these exscriptive methods produces the interior as a workable domain by generating exteriorised image surrogates which map the body's interior into some form of 'virtual' space – a mirror space or photospace. Through this production of image surrogates the opaque interior can be opened out to therapeutic instrumentation, an exteriorised interior, a surgical space where living organs and tissues can be therapeutically reworked, removed or replaced.

The history of twentieth-century medical imaging, which has the Visible Human Project as its most recent innovation, is, for the most part, a series of elaborations on the exscriptive logic driving the x-ray image, elaborations conceived of as 'corrections' to the ambiguity of its visual field. In particular the VHP is the result of subsequent medical attempts to exscribe the bodily interior not as photographic analogue but rather as spatial homologue, and not as visual shadow but as digital substance, as a visual positivity free of the duplicitous play of transparency and perspective found in the x-ray. In particular the VHP is the anatomical correlate of tomographic methods of exscription – *tomos*, to cut, *graphos*, to record. As I commented in the previous chapter, the VHP *literally* cuts the figures used in its image production in the same spatial configuration as conventional tomographic methods – CT scans and MRI. In this sense it is an anatomical, dissection-based technique rather than an exscriptive one, a technique of the corpse which mimics and literalises an existing diagnostic imaging modality.

Tomographic imaging methods had their earliest development in the 1920s, both within and without medical circles. Cartwright and Goldfarb (1992) locate their beginnings in experiments with both photography and radiography, experiments intended to 'correct' the problematic perspective already commented upon in relation to the radiograph. The visual field of the radiographic image is a three-dimensional flatness which collapses the body's depth and volume onto two dimensions as a series of spectral superimpositions. This perspectivalism of the radiograph, the absence of a neutral position from which to x-ray any interior structure, was already marked as a problem for the interpretation of radiographic images in the earliest years of the x-ray. Daston and Galison (1992) summarise a critique of the artefactual nature of the radiographic image found in an early radiological atlas:[10]

> By 1905 ... it was clear that there were systematic mismatches between macroscopic anatomy and the x-ray image of the human body. There were elements of the body which did not produce image traces on the x-ray, and there were representational elements on the x-ray that did not correspond

to identifiable characteristics under the anatomist's knife. As a result the diagnostician had to learn – through the study of an atlas such as this – to qualify the mechanical procedure of x-raying with a knowledge of systematic deviations between anatomy and its [radiographic] representative. Secondly the atlas could ... help to prepare the observer for the enormous variation of image that resulted from the movement of the x-ray tube or a rotation of the body part under scrutiny. Such a displacement of the body part could easily make certain contours disappear and other ones appear. Third, by collapsing a complex, three-dimensional form into two dimensions, the projective process itself could easily mislead.

(Daston and Galison 1992: 106)

Tomographic imaging is concerned precisely to eliminate the collapsed depth of the radiographic image, to try to banish the ambiguities of transparent density from the image by making its picture plane as shallow as possible. The CT scan, for example, eliminates all information outside the 1 mm depth of the scan, so that problems of perspective are minimised and the image field conforms as closely as possible to the projective medium, the screen. In this it shares a general drive within medical modernism to eliminate depth from both its image modalities and its objects, a drive which Cartwright describes as cubist – 'a [medical] culture that reconfigures the bodily interior as an endlessly divisible series of flat surfaces and mobile networks' (Cartwright 1995: 91). At the same time, tomographic methods paradoxically reinscribe depth in the image, or more precisely infer a composite image which *reproduces depth within itself*. Tomographic methods image the body as a registered series of planar 'cuts', which can, like the VHP data, be restacked to reproduce a volumetric map. Cartwright and Goldfarb comment on the shift from lenticular to planar space that took place in early tomographic experiments:

> [Tomography attempts] to map the space of the body in planar projections, each plane constituted as the sum of a multiple set of geometric projections ... the extreme reduction of the focal field counteracts the spatial imprecisions introduced by standard photographic perspective. It can be argued that [tomography] does not simply correct perspectival space, but introduces an entirely new model of spatial representation. The [tomographic] image is less a pictorial record of appearances than a means of measuring an abstract, non-visible property of the body: volume. ... Ultimately the individual images ... would be laminated so as to form a photographic block proportionate to the object. This dimensional model is important not because it reproduces the object, but because it facilitates the precise measurement of, and location of points within, a space that cannot otherwise be seen or photographed from any single vantage point.
>
> (Cartwright and Goldfarb 1992: 193)

Tomography paradoxically produces a duplication of volume through the elimination of depth 'computed as an absence, a non-visible differential between two images, two positions' (Cartwright and Goldfarb 1992: 196). It is an imaging modality which exscribes the interior not through a collapsing of boundary with depth but through a preservation of relations between boundary and depth treated as a spatial array, volumetric contiguity treated as a distribution of infinite points in space. Tomographic methods project the bodily interior as a precise spatial homologue. Hence they lend themselves to much more precise navigation within the body, locating organs and structure in a grid of anatomical morphology, a grid through which surgical trajectories can be plotted and forged.

By televising the endosoma, exscribing their traces as externalised image-objects, each of these imaging practices lends the bodily interior to shifting repertoires of instrumentation and work it into new therapeutic, biotechnical ensembles. Each development in technique constitutes a new attempt to deal with the particular material properties and morphological complexity of the living body. Exscription fractures the body's opacity and self-enclosure, so that in the process of being visualised its recessive depths become, simultaneously, remote projections. Its surface is broken up and confused, a surface which is only provisional and which can just as readily provide access to instrumentation. Each new technique tries to negotiate the convolution of organs in a way which both preserves their viability and vitality and lends them to further orders of instrumentation. Exscriptive medical instruments allow the body to be staged as both refractory volume and refractive image space, so providing surgeons with leverage and orientation in the task of moving *through* its morphology. Shifts in the image space, its particular aesthetic and metrical qualities, necessarily have implications for the ways in which the bodily interior can be spatially addressed.

The Visible Human Project is, as I have already commented, an anatomical form of a tomographic exscription. It is a writing out of the bodily interior which takes place in the same spatial orientation as optical tomography, but which substitutes an actual incision for the optical 'slice' of the CT scan. Its particular mode of exscription, the cryomacrotoming technique which planes and digitally photographs the frozen corpse, produces a set of image data which has unique exscriptive qualities.[11] While conventional tomography produces CT and MRI homologues of living bodies, only the anatomical technique of the VHP can produce a photorealistic tomograph, imaged in light spectra, in addition to the other two modalities. The three modalities are precisely co-registered to each other, so that the VH data can be used to test out and compare imaging research across the three media in a stable relation.[12] (See Figure 4.4.)

It is this co-registration of diagnostic media (CT and MRI) with an anatomic medium (digital photography) and the very high density of visual information conveyed by the VH data that make them so useful as both

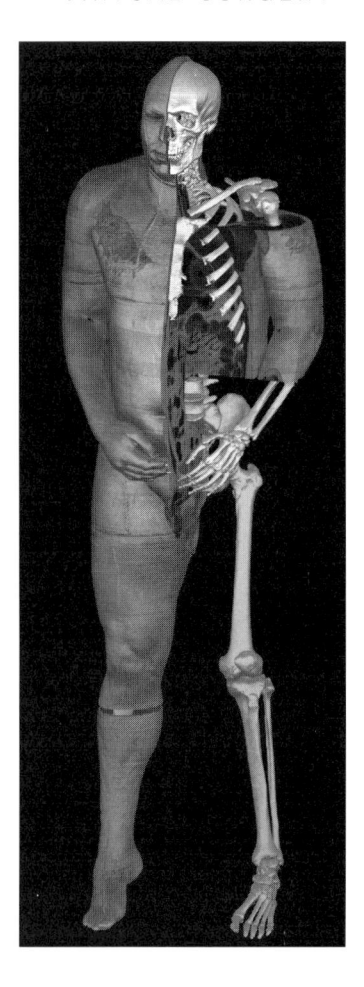

Figure 4.4 Three-dimensional reconstruction of the Visible Human Male, showing skin, internal organs and skeletal structure (image courtesy of William Katz, Varian Medical Systems)

speculative, surrogate objects and regulatory anatomical norms. The dataset's density, its photorealistic verisimilitude and its nature as *digital* visual data lends it the capacity to act as both substitute and supplement *for* living bodies within the terms of computer imaging research. It can act both as experimental test object, and as visual information which adds to and normalises the relatively meagre visual information able to be extracted from the living body, as distinct from the rich visuality made available by the corpse.

In what follows I want to consider in more detail these two orders of use: the dataset's usefulness as a surrogate, experimental object, and its use as a supplement for the information gatherable from living, rather than dead, bodies. Each of these uses have implications for ways that living bodies can be spatially addressed in both diagnosis and in surgery.

SURROGACY AND MORPH-OLOGY

It is evident from the applications already described that the VH data's capacity for surrogacy is being well used. The Visible Human figures are providing test-data for prototype development of new imaging modalities, particularly virtual endoscopy, and for the modelling of surgical procedures, among other uses. Parts of the data can be used to provide virtual knees, virtual eyes, virtual brains and skulls, which can be manipulated, deformed, dissected and ressected both visually and haptically. The deployment of the Visible Human data as these surrogate objects is, however, in no sense a use unique to these data. Rather, visual surrogacy is a central technique in biomedical research and in scientific research more generally. As Latour (1990) suggests, complete living entities themselves are, in the first instance, too obdurate and various for direct use in research and speculation. Instead, experimental science finds its appropriate object in the trace systems which are abstracted from them. The object of the experimental gaze is not the living entity itself but its visual surrogate, located in a generalised mathematical space of inscription. Systematic knowledge only becomes possible when scientists avert their eyes from the natural object and instead engage with the space of inscription.

> Scientists start seeing something once they stop looking at nature and look exclusively and obsessively at prints and flat inscriptions. In the debates [in the history of science] around perception, what is always forgotten is this simple drift from watching confusing three-dimensional objects, to inspecting two-dimensional images which have been *made less confusing*. ... The 'objects' are discarded or often absent from laboratories. Bleeding and screaming rats are quickly dispatched. What is extracted from them is a tiny set of figures.
>
> (Latour 1990: 39)

Such images and data do the crucial work of acting as standardised, co-operative, and simplified surrogates for the objects under consideration, while at the same time maintaining a specifiable, mathematical relationship to the objects from which they have been abstracted. Daston and Galison's (1992) term for these surrogates is 'working objects':

> Working objects can be atlas images, type specimens, or laboratory processes – any manageable, communal representatives of the sector of nature under investigation. No science can do without such standardised

working objects, for unrefined natural objects are too quirkily particular to cooperate in generalisations and comparisons.

(Daston and Galison 1992: 85)

Working objects are objects in process, which can be used experimentally to test out certain morphological and biotechnical propositions. The Visible Human dataset is a particularly information-rich, flexible working object and visual surrogate, with dynamic modelling capacities which lend it to a whole repertoire of experimentation. It acts as an *aide-pensée*, a hypothetical model which sets out and enables certain kinds of thought, experimentation, visual projection and fantasy about the operation of the human body and its possibilities for transformation. It facilitates certain kinds of imaginative theorisation about possible ways to technically address living bodies and bring them within the forms of visual calibration made available by computer vision. Such theorisation is central to the process of producing new kinds of medical knowledge, new repertoires of technique and new explanatory systems.

At the same time the Visible Human dataset's usefulness for such speculation is not simply free and unimpeded. Rather, its speculative power as a biomedical research object depends upon its conditioning by the material particularities of the living bodily interior, and by the material qualities of the image itself, its particular mediatic properties which lay out certain kinds of spatial and aesthetic possibilities and pre-empt others. The kinds of speculation made possible by the Visible Human data depends then on the dataset's status as a compromise formation which literally incorporates a knowledge of bodily ordering, the exscriptive homologue produced by the cryomacrotome, into the digital medium of the computer. The Visible Human Project hybridises body and computer, splicing together two orders of matter, flesh made data and data made corporeal, so that the image retains some of the properties of both.

In doing so it creates a parallel space of anatomical and physiological demonstration which is less confusing and more orderly than any living anatomy could be. As both Latour and Daston and Galison emphasise, the usefulness of 'working objects' for science derives from their status as objects within closed systems. They are located within laboratory systems or visual systems which have a degree of internal consistency and completeness absent from the open system which characterises relations between living body and world. Hence they can be made to undergo certain kinds of uses, certain kinds of transformations (the laboratory experiment for example) under controlled conditions, transformations which are amenable to standarisation and hence able to generate stable, general laws about the material organisation of the world (Latour 1993).

The Visible Human dataset works as a speculative, operative image precisely because it incorporates so much information in the sense of positive anatomical content, yet it excludes the kind of complexity which makes the multiplicity of lived bodies unpredictable and uncooperative, open-ended and incomplete. This peculiar capacity for complexity within closure derives from its status as simultaneously anatomical trace and digital code. The Visible Human archives are material residues of once existing bodies whose matter is now visual data, mathematical code, whose forms take place within the formal, mathematical rules of data space. According to Robins the world of the virtual screen

> is made tractable and composable ... a logical universe of [simulation]. ... The [virtual world] is a safe and predictable environment. Like other kinds of micro-environment (board games or card games, for example) it is structured by a set of rules, a set of assumptions and constraints, and within the terms of these rules everything is possible, though nothing can be arbitrary or contingent.
>
> (Robins 1996: 48–9)

The Visible Human data are literally operable images, images which can be manipulated as if they were organs in a surgical operation, because they partake simultaneously of the qualities of photographically based anatomic trace and mathematical code. As photographic traces they can claim the verisimilitude of the photograph, its indexical status as record of an anatomical body. Hence the VH data can underwrite their surrogate status through an appeal to photographic truth value, as exscriptions of organs which can be substituted for organs themselves. At the same time they can be worked over in all the ways specific to the image economy of visual data and the domain of simulation.

As digital photographs, the VH data exist not as continuous field of shade variation, the non-quantifiable, analogue surface of seamless gradations which is the photograph, but rather as a discrete grid of image elements. The digital camera used in the cryosectioning procedure works, as do the CT and MRI scans, through the allocation of a set of quantitative values (integers) to different qualities of radiation. In the case of the CT scan, values are allocated to x-ray radiation, according to its degree of attenuation by different tissues. In the case of MRI values are allocated to differentials in electrical emission of the body's hydrogen nuclei. In the case of the digital camera values are allocated to intensities of colour, lightness, darkness and texture – the visual spectra associated with conventional photographic vision. The resulting array of integers can be stored, transmitted, copied and altered by the computer, just as any array of mathematical information can be treated. At the same time each integer specifies a pixel (picture element), the light unit of the virtual screen, which composes the virtual image as a finite, discrete grid of light points through visual interface software. As a consequence, the digital photographic image displays qualities of mutability far in excess of mechanical photography or cinematography.[13]

> Photography and digital imaging diverge strikingly, for the stored array of integers has none of the fragility and recalcitrance of the photograph's emulsion-coated surface ... the essential characteristic of digital information is that it can be manipulated easily and very rapidly by computer. It is simply a matter of substituting new digits for old.
>
> (Mitchell 1992: 7)

As mathematical data the Visible Human images operate within the closed, logical space of mathematical calculation. Their mutability is algorithmic, effected through the application of formulae. It is this quality which makes them morphable morphologies, volumetric traces which are amenable to the deformations and reformulations that characterise visual data objects. Like all digital images they have the quality of limitless mutability without loss of information. Moreover, in medical imaging based on MRI and CT scans, the value of pixels or voxels also conveys information about tissue type. In CT different intensity ranges indicate soft tissue or bone, while MRI intensities can be used to discern a shifting range of physiological as well as anatomical features (Pommert *et al.* 1993). These imaging methods consequently give a mathematical precision to the differentiation of tissue types, producing images that demarcate tissue at the level of discreet pixels and voxels, making an anatomical model with clearly marked volumes and structures, and absolute divisions between contiguous organs and tissues.

These capacities give the VHP figures their attractiveness as operable working objects. While as meticulously registered anatomic records they provide an indexical, perceptual anatomy, as morphable objects they also provide a propositional anatomy; they are simulated image objects which can be used to interrogate and extend the limits of the living body's macroanatomical transformations and accessibility to visual and surgical instrumentation.

THE FLYTHROUGH

All of the applications for the VH data described at the beginning of this chapter draw to some extent on these particular speculative, mediatic qualities, the algorithmic performativity of digital data in general. The data application which, it seems to me, makes the most *exhaustive* use of the VHD's particular mode of exscription – its photographic tomography, the algorithmic ordering of data, its display of tissue typing, and its propositional qualities – is the flythrough sequence. I have already described the visual logic of the flythrough to some extent. It is a way of rendering the data so that the user's point of view moves through the body as a miniaturised, airborne point, which flies the length of various tubular anatomical structures. The flythrough is effectively the point of view adopted for all virtual endoscopy. As I described earlier, the Visible Human data are currently being used as test-data for the development and standardisation of virtual endoscopy as a diagnostic method and as an adjunct in surgical procedures. Flythroughs have been and are being constructed which traverse the same organs as conventional endoscopy – the oesophagus, the stomach, the colon, the fallopian tubes and the like. Flythroughs are also being constructed for organs where conventional endoscopy is difficult or unobtainable: the inner ear, the heart and the spinal canal for example (Robb 1996). The Visible Human data are being used in these cases as prototype data, for the testing out and refinement of imaging techniques, rendering algorithms and the like, before application to patient data, and also as a control against which patient data can be judged.

Virtual endoscopy substitutes data gathered through tomographic methods for data gathered through the use of a conventional endoscope. The conventional endoscope involves the introduction of the imaging technology directly into the lumen itself, the insertion of cameras into the internal length of the organ. The tomograph on the other hand produces data about the interior surface of organs by virtue of its cross-sectional access to all points of the body's depth and length. Flythroughs mimic the point of view of conventional endoscopy to some extent, its traversal of the interior surface length of the lumen, its forward orientation. Nevertheless it is not an indexical account of the lumen's traversal, a recording procedure, but rather the simulation of such a traversal. The adoption of simulation and modelling as a means of knowledge is not of course restricted to the Visible Human data, but has become general to the application of computers in all domains of science and technology. Simulated scenes have become crucial modes of scientific imaging in astronomy as well as medicine, geology and other sciences. As one observer describes scientific simulation, 'the procedure is to employ some appropriate scientific instrument to collect measurements and then to construct perspective views showing what would be seen if it were, in fact, possible to observe from certain specified viewpoints' (Mitchell 1992: 119).

Flythroughs are constructed algorithmically, through mathematical programming procedures that project a traversable volumetric space from the flat, serial data files obtained from the cryosectioning process. A scientist describes the development of the virtual colonoscopy at SUNY in the following terms:

> 3D virtual colonoscopy ... utilises computer graphics technology to visualise the inner surface of the colon based on CT images of the abdomen. ...
> A set of newly developed visualisation algorithms has been applied on acquired CT data ... as well as the Visible Male data. Our algorithms

VIRTUAL SURGERY

> support both fully-automated flythroughs and interactive walk-throughs along the inside of the colon. To generate a flythrough automatically, we employed an algorithm which computes a flight path while guaranteeing that the virtual camera always stays in the centre of the colon to achieve better viewing of the colonic surface.
>
> (Hong *et al*. 1996: 1)

Medical flythrough software utilises and adapts existing flight simulation and flight path planning technology (Meglan 1996) to project traversable volumetric interiors out of two-dimensional data. Like the terrain in flight simulations the Visible Human data comprise a self-enclosed space with its own internal logic of perspective and traversable sequence, made visible from a point of view which looks 'inside out' at the interior surfaces of organs.

This point of view is described as endoscopic, and it is being used to traverse the same organ paths as conventional endoscopy. Nevertheless the point of view obtained in the former is not a simple 'improvement' on the latter, but is effectively a point of view impossible to obtain through conventional endoscopy. The conventional endoscopic view involves the optical fibre camera pushing through and forcing open the self-enclosure of the lumen, so that its point-of-view is rarely unobscured by an immediacy of enclosing mucosal surface. The endoscope cannot view the interior walls of lumen structures from a 'free' position. Rather it is always caught up in the self-enclosed restriction of the lumen's musculature and often in a surface-to-surface juxtaposition of lens with mucosa. The depth of field in a conventional endoscopic image is hence extremely shallow or nonexistent. Its images are saturated in glistening bloody secretions, occluded by the pulsation and muscular self-enclosure of the organ examined, and spatially ambiguous to an untrained eye, giving no clear sense of directionality or orientation. The conventional endoscopic point of view is always conditioned *in relation to* the configuration of flesh through which it moves, even as it reconfigures that flesh by its presence.

The virtual endoscopic flythrough view, on the other hand, operates as if its presence has no effect on the terrain through which it moves, and as if it itself has no dimension. It is a perfectly disembodied and frictionless point of view which can travel at any speed, unlike conventional endoscopy whose view is demonstrably enmeshed in a particular set of instrumental dimensions and bodily recalcitrance. The virtual endoscope acts as if the anatomical terrain is completely open to aerial traversals, in fact as if it invites such traversals. It presents a dry, smooth-surfaced and completely hollowed lumen space, free of the cloying strictures and ambiguity of the conventional endoscope. The interior surface of the lumen is fully available to simulated vision, appearing as perspectival surface with an infinitely converging depth of field, unlike the non-perspectival, cloachal space of the conventional endoscope. The flythrough view conveys a strong sense of the body interior as extended vacuum traversed by clearly demarcated and solid structures (see Figure 4.5), a space which lends itself to the plotting of flight paths and heroic aerial traversals.

Unlike the qualitative, continuous anatomic domain encountered by the conventional endoscope, the virtual endoscope addresses an anatomic domain which is completely calculable, because it is already gridded quanta, a mathematical description of the body laid out as geometric space. Hence it is open to detailed specification of relationships between point of view and data, and to determinations of location and orientation, just as a pilot in a flight simulation can accurately determine virtual position in simulated space. Robb describes these qualities:

> Virtual endoscopy provides viewing control and options that are not possible with real endoscopy, such as direction and angle of view, scale of view, immediate translocation to new views, lighting and measurement.

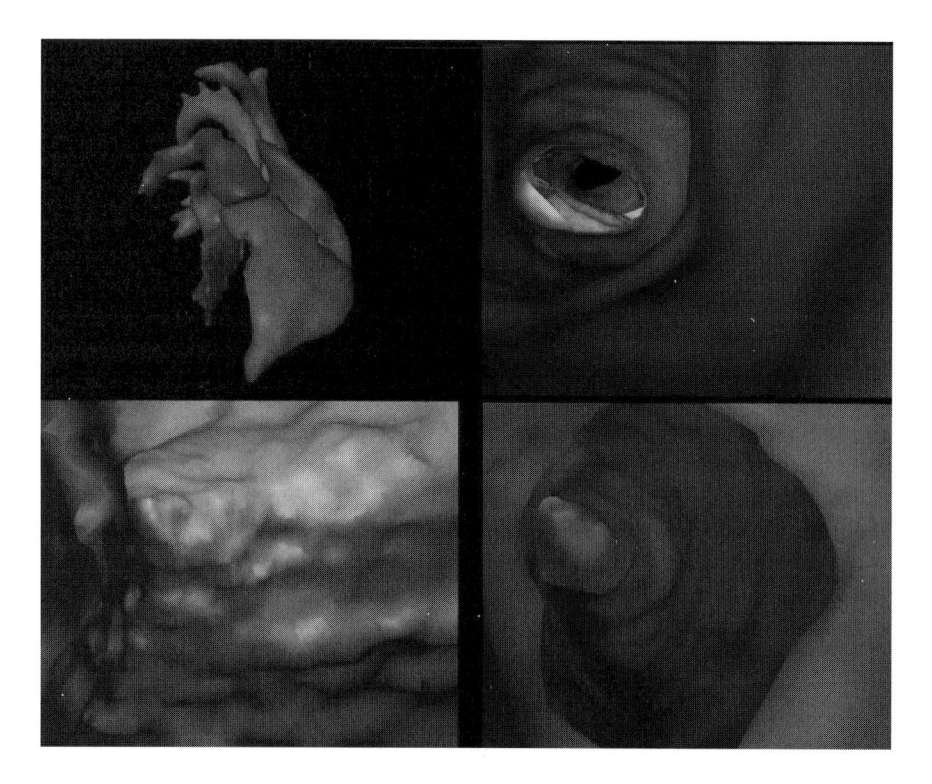

Figure 4.5 Heart model computed from Visible Male dataset, and three virtual endoscopic views showing interior surfaces of major vessels and heart chambers (image courtesy of R. Robb, Mayo Foundation)

> Visual feedback positioning systems and navigation guides can orient the virtual endoscopist relative to the actual anatomy. Image values from the original data can be associated with any view and quantitatively assessed.
>
> (Robb 1996: 2)

The screen can juxtapose two or more points of view simultaneously, providing both immersive views and overview orientation images of the data which show the location of the endoscope (Lorensen *et al*. 1995). Moreover, unlike conventional endoscopy, virtual endoscopy can move laterally to parallel structures within the body. Camera-based endoscopy is necessarily caught within the trajectory defined by the morphology of the lumen itself. It cannot shift from the inside to the outside of the organ, nor can it move to other adjacent organs. Virtual endoscopy has equal access to all interior and exterior surfaces, and its trajectory is determined not by the morphology of organs but by the orderings of simulation and hyperspatiality.

The flythrough is a mode of exscription that maps the body according to the rational space of cybermetric simulation. As working object it frees up and simplifies the ambiguous and restricted point of view presented by the endoscope, which encounters the bodily interior as a series of heterotopic, disjunctive blind points and obdurate surfaces. The virtual endoscope, enabling the clear visualisation of pathological structures and the traversal of clean interior spaces, is an intelligible and navigable writing of the anatomical body which recapitulates the crisp aesthetics of the two-dimensional anatomical text and transforms a series of heterotopias into a homogeneous, utopic space. Virtual endoscopy enables the entire body to be visualised using the inside-out point of view associated with endoscopy. It does not require any point of access into the lumen space, but can rather view the interior surface of all organs equally. Above all, it simplifies the morphological self-contiguity of the body, rendering its yielding tissular density as an architectonic structure. Its ability to model densitometric information allows it to present clear demarcations in tissue type, enhancing its simplification of tissular and functional relations. The virtual flythrough lends the interior of the body to the mathematicised space of the scientific endeavour, and weakens its resemblance to the grotesque, involuted non-space of abjection which appears through the endoscope. In its precise calibration of anatomical space, where each pixel or voxel is at least potentially specifiable as a locale in the body, it grids up the body as an array of infinitely calculable, discrete and locatable points, both infinitely composite and decomposite. Hence it makes the body potentially available to the most precise kinds of clinical diagnoses, surgical navigation and interventions, the excision of tumours for example, where precise demarcations of depth, location, tissue type and tumour morphology are essential to the success of the operation. The flythrough schema facilitates this precision because all points can be precisely plotted in relation to each other, and each point has a specifiable address, a co-ordinate which can be co-ordinated in relation to other orders of instrumentation.

MORPH-OLOGY AND THE ANATOMICAL NORM

The flythrough is a mode of exscription and projection, but like all of medicine's working objects its projection is not simply a matter of mathematical extrapolation. Rather, it projects in a double sense: it produces both morphological projections that extrapolate data, and screen icons, which embody medicine's desire for bodies that are utterly accessible to vision, navigation and instrumental intervention. As already quantified bodies, they offer no resistance to manipulation in rational, homogeneous space. The Visible Human flythrough is simultaneously a practice and a fantasy of mathematicisation, but this is not to suggest that such fantasisation excludes clinical applications. As I suggested earlier, science's surrogate objects retain specifiable relationships with those things which they trace, so that their experimental deployment can work recursively on the things for which they substitute.

Hence the Visible Human data flythrough describes the visual logic of some new uses being made of patient data during actual surgical procedures, and acts as a supplementation and regulatory norm for such patient data. Current surgical research involves the use of patient data reformulated as homological models. I have already described ways that such data homologues are used in surgical planning and rehearsal, but some emerging forms of surgical practice also use data homologues *intraoperatively*, as ways to enhance the intelligibility and navigability of the patient's body during surgery itself. So, for example, 3D patient-data models are being used at the New York University School of Medicine as part of surgical procedure, to help navigate and excise brain tumours. Surgeons use a virtual display which shows the surgical patient data as an operable model. Surgical instruments are tracked and displayed on the screen in relation to the patient data. For the surgeons this would have multiple advantages:

> These techniques would provide a view of the brain tumour, which is virtually indistinguishable from surrounding healthy brain tissue in the body, in stereoscopic 3D; they would also show its relationship to important landmark body structures like the skull, facial features, brain surface, veins and arteries. Finally they would show the proximity of the surgeon's instruments to the tumour.
>
> (Siegel 1995: 59)

Intraoperative visualisation enables tissues to be displayed as locales rather than semi-amorphous or undifferentiated masses, which is how they might appear in direct visual contact, and guide the surgical intervention to avoid collateral tissue damage. This is one of the uses of virtual endoscopy, as a way to visualise prostate cancer for example.

> Virtual endoscopy can be performed on these patient specific models to precisely visualise the interior of the prostate and embedded tumor relative to critical anatomic structures, such as the urinary sphincter, seminal vesicles and neuro-vascular bundles, helping to define appropriate surgical margins in order to reduce the risk of post-operative complications.
>
> (Robb 1996: 7)

The promise of such patient homologues is that they can render patient data with the kind of precise availability to vision and instrumentation promised by the VH flythrough. The patient interior can, it is hoped, be envisaged as a precisely calibrated terrain, a homogeneous and continuous mathematical space which lays out pathology as clearly demarcated features in an intelligible landscape, where each organ and tissue is specified on the xyz axis of geometry. The refinement of such homologues would extend the possibilities of keyhole surgery and telesurgery, because the data can be used to indicate and guide the location of instruments. Visual feedback positioning systems and navigation guides can be utilised with the data homologue, showing surgical instruments in the patient's interior as points on the homologue.

Nevertheless, living bodies have proved recalcitrant to a seamless assimilation into data homologues and arraignment in digital space. As I noted earlier, the Visible Human data are so rich in part because they have been abstracted from corpses. As sacrificial bodies these corpses could be analysed according to a maximum visuality which the living body does not permit. Hence patient data are, by their nature, more discontinuous and fragmented than the rhetoric of virtual medicine would generally allow. While CT and MRI scans are described as 'non-invasive' they nevertheless involve the systematic traversal and alteration of the body's tissues, and safety standards require a limitation to the density and intensity of these imaging methods in clinical applications. As one scientist describes these limitations:

> In order to achieve these [virtual] applications ... the models we extract from the data must be accurate and meaningful. Accuracy is compromised due to the fact that digitised scans are samplings of some aspect of the body's tissues, and are not continuous in nature. To further complicate matters, scans are taken at intervals through the third dimension which usually greatly exceed the intervals between data samplings in the plane of the image. Decreasing the intervals between images means creating greater discomfort or danger for the patient. So we must interpolate between the existing data, introducing inaccuracy.
>
> (Siegel 1995: 59)

The formulation of patient data therefore always involves speculation and projection, and one of the major uses for the Visible Human data is as a tool for refining methods of interpolating patient data into the idealised mathematical space of digital imaging. The VH data are being used as digital norms to develop algorithmic methodologies for volume visualisation which can be applied to all patient data. It will provide standardised ways of extrapolating that data into virtual 3D formats, so that it more closely resembles the seamless, rational space of the VH flythrough itself. Here the biovalue produced by the VH data derives from what I would term its biosemiotic productivity. Just as the VH data can itself be repaired and reformulated again and again, so too can it be used to repair the discontinuities of patient data. Tomography works by breaking up the contiguous volume of the body into a series of cross-sections, but the dream of virtual tomography is to reformulate a body homologue without loss, a point-to-point indexicality between analogue body and digital duplicate. As a mathematical description of corporeal space much more information dense than the descriptions made possible by scans of living bodies, it can be deployed to smooth out the latter's disjunctures, extrapolate a contiguous and homogeneous anatomical space from the discontinuous arrays derived from the patient's mortal body.

OPERATIVE IMAGES

The term 'archive', at least in its everyday usage, implies a system of storage and retrieval for a specific content, which remains untouched by the form of its containment. In analysing the bibliometric and cybermetric archives of human anatomy in this and the preceding chapter, I have, on the contrary, demonstrated the extent to which container and contained lack any specifiable limit. Rather, the content of the human is hopelessly interramified with that body of techniques which purports to merely analyse and communicate it. If anatomy is the oldest 'science of Man', as Foucault (1975) suggests, it is also the beginning of a relentless experimentation with ways that the material properties of human bodies, living or dead, can be engineered in relation to the material properties of writing technologies, forms of inscription and exscription which produce the body as a system of biographs. Such experimentation has, to use a Latourian term, allowed 'a swapping of properties', so that the body has been made accessible to some of the kinds of intelligibility and navigability particular to the book and the computer, and the book and the computer have been able to embody the human. At the same time this interramification undermines any attempt to position the human body as a pretechnical, natural point of origin for science's accounts of it. Such accounts can only demonstrate forms of processual *relationship* between science's own techniques and particular bodily qualities and states.

Anatomical images, then, are particular kinds of compromise formations, setting out the terms of translation through which body and archive are confused. In this sense they are 'operative' images. They exercise certain kinds of agency in biomedicine's attempts to materially order the living body, setting out morphological and biotechnical propositions and possibilities through their particular performative and mediatic qualities. While he does not himself use the term, Latour suggests the significance of such operative images in more general terms when he rejects the passive tool status usually attributed to technology in conventional sociological studies, and the location of all agency and mastery on the side of the human. He argues for the necessity to

> redistribute essence to all the entities that make up [a technological] history. But then [technologies] stop being simple, more or less faithful, intermediaries. They become mediators – that is, actors endowed with the capacity to translate what they transport, to redefine it, [and] redeploy it.
> (Latour 1993: 81)

The Visible Human data are operative images to the extent that they set up living bodies so that they become traversable and operable according to the spatial logics of computer vision. The sacrifice of two bodies in relation to the matrix of computer vision produces methods whereby other, living bodies can be brought into productive, therapeutic orientation to that matrix. As operative images, the Visible Human data create forms of translation that facilitate the enrolment of bodies within networks of computation, so that morphology becomes morphable, workable within the terms of virtual vision and simulation. It is the creation of such operative images, and other biomedical applications of the computer, which has made the computer a primary technology of life.

5

IATROGENESIS
Digital Eden and the reproduction of life

> Man makes man in his own image
> (Norbert Weiner, 1968)

The Visible Human Project deploys the computer's potential as a technique for the production and management of life and biovalue. So far, only one mode of its deployment has been considered – the forms of medical management specific to the body at the scale of macro-anatomy. The fundamental biomedical problem of the body's opacity and self-contiguous depth, the obstacles that its morphology throws up to visualisation and traversal, and the kinds of solutions presented by the Visible Human Project and its cognate forms of diagnostic and anatomical imaging, have been investigated at length. The computer's capacities for complex, animate visuality, and its ability to render all data as visual data, lend it to the development of a new repertoire of visceral prosthetics, new ways to address the bodily interior as exteriorised assemblages. In the detection and treatment of cancers, tumours, internal lesions and organ anomalies, and in its specification of anatomical norms, the VHP and its cognates work to intensify, preserve and manage health and vivacity along particular, biomedical lines.

If the computer has become a technology of vivification, the implications of this development nevertheless exceed the kinds of immediate biomedical applications so far examined. All of medicine's biotechnical innovations have implications for vivacity, health and illness, and many enter into and alter the relationships between life and death. The computer, as the new meta-technology of life, has participated in some major reconfigurations of this relationship. Through the auspices of computation and molecular biology, the qualities of vivacity have been redistributed throughout systems of information. The force of life is increasingly posed and manipulated as bioinformation, a negentropic programming of matter which lives as replicating, mutating code. Casting life along such lines necessarily alters the terms of death as well, although in general death remains unthought in bioinformatics, presented simply as a *lack* of information.

In the two chapters which follow, I lay out some of the implications of this bioinformatic logic, as it plays itself out through the aura of uncanny animation which surrounds the Visible Human Project. The project's claim to model the *living* body depends on the more general vivification of information, the equation of the screen space with a saturated space of negentropy. Nevertheless, the deployment of corpses as the raw material for the project's images necessarily complicates any attempt to locate the computer screen as a space of untroubled vivacity, so that the Visible Human Project tests out some of the asymmetrical redistributions of life and death effected by the life of information.

DIGITAL EDEN

One of the more insistent rhetorics about the Project, deployed by both the popular media and the project scientists themselves, is its likening to a new Genesis, and the nomination of the figures as a new Adam and Eve. Headlines in *Fortune* magazine, *Der Spiegel*, *The Washington Post* and *The Economist* all used Genesis puns of various kinds to describe the project. This tendency was particularly evident at the launch in late 1995 of the Visible Woman data, which was almost universally represented as the provision of a mate for the Visible Man, an Eve sent to cyberspace to provide companionship for a solitary Adam.[1] Media stories featured headlines like 'Virtual Eve joins Virtual Adam', 'A Digital Adam' and 'Adam's Family Values'.[2] Articles typically use the Genesis story to comment on the gendered order of the bodies' publication on the Web. One article states, 'First came the "Visible Man," the Adam of cyberspace, a human cadaver transformed into billions of bits of digital information and sent along on the internet. Now, one year later, comes the "Visible Woman," the Eve of cyberspace' (McNamee: 1994). Another story speculates that 'At some future date, Adam and Eve may have progeny ... the NLM would like to add imaged cadavers of children' (Stix 1993: 123). The idea of a 'virtual family' had some appeal to the media as an extension of the Genesis reference; a journalist at *The Economist*, anticipating the possibility of Visible Embryos and Visible Children refers to 'Adam, Eve and little Cain'.[3] If this seems like journalistic licence, a resort to sensational metaphor in order to make science more palatable for a popular readership, it should be noted that the nicknames Adam and Eve were in use *within* the project even before the actual bodies were located and imaged (Stix 1993). More recently the nicknames have had to be withdrawn, because a North American multimedia company has patented the name A.D.A.M.![4]

This rhetoric could easily be disregarded as simply another manifestation of scientific hubris, or as a resort to clichés of origin by an impoverished medical imaginary. Certainly the deployment of Genesis iconography is habitual in the history of medicine, and can be found at earlier points in the history of anatomy. The figures of Adam and Eve (see Figure 5.1) were repeatedly used in anatomy texts between the early sixteenth and the late seventeenth centuries to stand for divinely ordained physical norms:

> Artists would use the prototypes of Adam and Eve, perfect in health and immortal before their expulsion from the Garden, their pristine physical state reflecting their innocence before the Fall ... as God's supreme creations, their bodies represented anatomical perfection.
>
> (Cazort 1996: 37)

In more recent times Genesis iconography has appeared in the literature of genetics, particularly the Human Genome Project (Myers 1990) and in the field of Artificial Life research (Helmreich 1998). According to Myers this general purchase of Genesis iconography obtains through its historical usefulness as a means of rhetorically substituting biomedical accounts of the origins of life for theological ones. It helps shift the locus of origin away from the domain of God and towards that of a mechanically conceived nature, so that the specification of normative bodies now takes place under the sign of biomedical prosthetics rather than divine creation. In this contemporary context, Genesis iconography is pressed into service to designate the

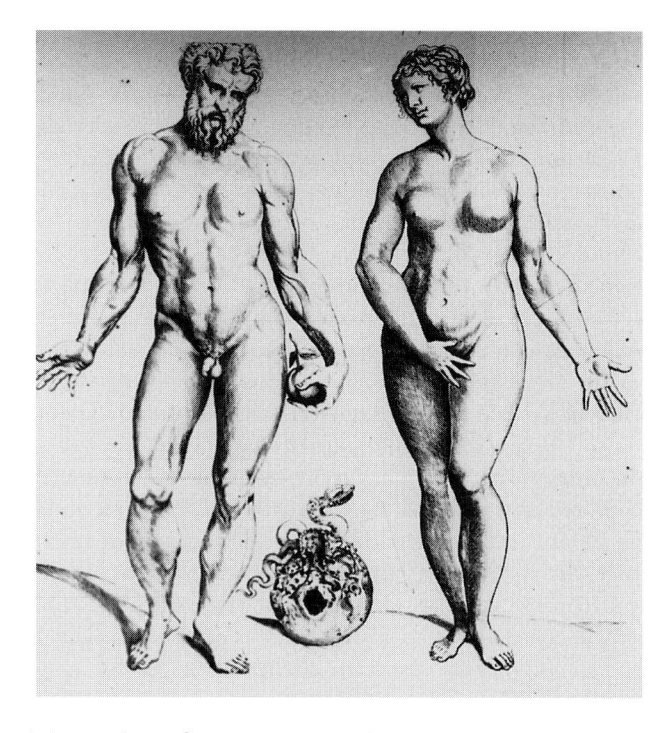

Figure 5.1   Adam and Eve figures, in *Compendiosa Totius Anatomie Delineatio* by Thomas Geminus, 1545

perfectability of 'Man' in the new Eden of biotechnical innovation.

If Genesis iconography is so generally dispersed, what use can I make of it here, in a more specific attempt to tease out the bio-politics and bio-semiotics of the VHP? While its use in this particular case no doubt furthers the metaphor's general function, it also articulates more specific claims. By paying attention to the particular permutations of Genesis iconography in this case I want to make it articulate specific relationships between user and screen, between nature and cybernature, and between 'original corpse' and 'copy' that obtain in the project. Above all I want to make it articulate what I term, rather ironically, 'IatroGenic desire'. This term is a wilful reconfiguration of the everyday meaning of iatrogenesis, which refers to secondary pathology created by medical intervention addressed to a primary disease. The *Oxford Medical Dictionary* defines iatrogenic illness as 'a condition that has resulted from treatment, as either an unforeseen or inevitable side effect'. Iatrogenic illness is thus disease which has its origin in biomedical procedures themselves.

In its usual sense, then, iatrogenesis is a term acknowledging that living bodies are not readily configured along simple axes of technical cause and effect. Until this point in the text I have stressed the body's openness to forms of technical redistribution and reconfiguration. Nevertheless, such openness does not imply stable, additive forms of technical embodiment. Prosthetic transformations – surgery, pharmaceuticals, artificial organs and limbs – involve the provocation of often unpredictable instabilities and losses, as well as therapeutic gains. Such transformations may involve a whole redistribution of the embodied subject's qualities, capacities, orientations and positioning in the world which do not necessarily produce a predictable outcome. So, for example, the recent development of antiretroviral therapies for people who are HIV positive has been successful in controlling viral replication, but it has also produced a range of disturbing side effects – redistributions of body fat, nausea, rashes – which disable and confront the user's sense of embodied identity and agency (Race 1999). The development of iatrogenic effects bears witness to the contingency of any relationship between specificatory technology and lived body, the activity of a mutable bodily field which does not simply receive and process a technical inscription in a sequence of cause and effect. Any biotechnical intervention inscribes itself into a complex, dynamic of corporeal animation and relationship, which redistributes its intended effects according to its own shifting logics.

'IatroGenic desire' is my term for a kind of reaction formation to this instability which can be located within medicine's imaginary. It is the desire to create, not disease, but rather kinds of bodies which are stable, self-identical entities rather than fields of perverse contingency. It is a product of the constitutive turbulence in any field of biomedical technique, the uncertainty of its embodiment, and a desire for living bodies which predictably embody such technique, which comply with medicine's fantasies of perfect management. IatroGenic desire is a kind of authorial desire in that it wants to 'make up' entities as acts of technical creation, through technically specifiable procedures which will produce reliable forms of life. These in turn can be reproduced in stable, technically manageable ways. Hence it is concerned with the marking of a moment of origin, a genetic moment for living entities produced through biotechnical procedures. While it may employ sexual reproduction, it is profoundly opposed to its vagaries, and attempts instead to substitute the reliability of commodity reproduction, the replication of the same.[5] As it circulates in the early twenty-first century, IatroGenic desire could be summarised as the desire for *programmable matter*, for a capacity to order materiality according to the algorithmic efficiencies of the computer.

The operation of this desire is evident in various domains of biology and biomedicine. One example can be found in the many reductive uses made of the idea of genetic code in biotechnologies, where the possibility of stable morphogenesis directed by a determining code script is proffered.[6] The claims of future prophylactic developments made on behalf of Human Genome Project depend on the possibility of producing healthier bodies through the technical manipulation of code, in a stable relationship of technical cause and bodily effect. Hence the mapping of the human genome undertaken by the project will, it is argued, produce, some time in the future, techniques for the replacing of defective genes with normal genes. This will, in turn, correct inappropriate morphogenesis or pre-empt the development of genetic predisposition diseases – a classification which has expanded rapidly to include many cancers, cardiovascular disease, neurological conditions and autoimmune diseases (Hood 1992). Critics of genetic determinism would point out that such diseases are strongly tied to environment, life circumstances and a complex range of other factors, which interact in unpredictable ways with the play of genetic code. The patenting of genetically engineered entities like the OncoMouse, on the grounds that they can be considered to be 'man made', and the susceptibility of such entities to the legal discourse of the patent and the commodity, is another development tied to the claimed power of genetic code to both produce and reiterate forms of life in a predictable way (Vasseleu 1996).

It could be objected that the operation of IatroGenic desire cannot be said to obtain in the case of the VHP, given that it is only an image, and not a living entity. My reply to this objection would be twofold. First, the VHP, by producing an animated, volumetric surrogate for a living body through the use of binary code, provides an iconography for this desire, a spectacular theatricalisation of its concerns which could count as a kind of IatroGenic pornography. The VHP is pornographic in the sense that it gives an exhaustively visualised form to IatroGenic desire, presenting a form of the human body which is perfectly repeatable and technically reproducible without loss or variation. Moreover it is a picturing of this desire which claims to leave no recess beyond the reach of the medical gaze, to show 'everything' in an excess of spectacle. The anatomies dismember and remember at will, give way to the relentless penetrative gaze of the flythrough, yield themselves up to any user as reproducible spectacle with infinite degrees of malleability. Every point in its space is available for visual use and consumption, no recess of the body icon is concealed or concealable. It is also pornographic in another sense, suggested by Braidotti (1994) in her treatment of medical representation. Medicine is pornographic, she writes, in that it deploys 'a system of representation that reinforces the commercial logic of the market economy. The whole body becomes a visual surface of changeable parts, offered as exchange objects' (Braidotti 1994: 25). That is, medical images in general, and the VHP in particular, partake of a certain commodity logic, where organs are treated as standardised, interchangeable, detachable use values. To this extent such forms of representation play out IatroGenic desires for precisely such regimes of standardisation and exchange.

My second response would be that the separation of images and technologies from living entities and what counts as life is never quite secure. The constitutive confusions, both within the biomedical imaginary and elsewhere, between 'life' and 'the illusion of life', and between the reproduction of an image and the reproduction of life, will be discussed below. At this point my claim is that insofar as the VHP is posited as a kind of artificial life, it acts out IatroGenic desire. In making this claim I am somewhat at odds with Kember's (1998) diagnosis of the VHP, when she speculates that it is a symptom of medicine's desire to 'father itself', to engage in autonomous acts of reproduction which go forward without recourse to the complications of the maternal. She writes:

> I want to argue that the prospect of a fully autonomous reproductive science and technology is a patriarchal fantasy which is enacted or acted out to an inevitably limited degree in the face of women's persistent ability to have children. ... The fantasy of fathering offspring without women is a defense against the anxiety provoked by the ultimate limitation placed on this desire. In the same way I will ... argue that the recreation of Adam and Eve in cyberspace is also an omnipotence fantasy enacted in the face of medicine's generative limitations and by means of a fetishistic use of technology.
>
> (Kember 1998: 88)

In making this argument Kember is drawing on a well established feminist critique of biomedicine and biotechnology, a critique developed primarily in relation to the desire informing reproductive technologies.[7] My first point of departure from Kember's argument turns on the implication in her analysis that for biomedicine the VHP represents an act of autonomous creation, a reinstantiation of the moment when God made the world and Adam *ex nihilo*. Such a claim may well represent the final expression of IatroGenic desire. Nevertheless, my reading of the VHP addresses itself to constraints placed on any idea of autonomous production by the necessity to defer to a putative 'original' body, the fleshly bodies processed for the project, and to manage the interpretation of the VHP as a 'copy'. The genesis involved in the VHP's IatroGenesis implicates the temporality not of creation but of invention, a second order originality which ties it more closely to the commodity than to parthenogenesis. In making this distinction I would also mark a second point of departure from the argument referred to in Kember's analysis. By regarding biomedical technologies as narcissistic tools of masculine self-creation, these analyses have neglected the extent to which the technologies they discuss are additionally or alternatively explicable through the logic of commodity (re)production applied to living beings, indicative of a desire not so much for self-reproduction as for stabilisable forms of multiple iteration, the standardisation of life. In other words it seems to me that IatroGenic desire is more an effect of certain machine logics than it is an effect of phallocentric psychology, although phallic narcissism may find pleasure in the kind of mastery that IatroGenesis implies.

THE ANIMATION OF VITALITY

To pose the data figures imaged in the VHP as Adam and Eve is to specify a particular imaginary matrix of relations between the project's scientists, the nature of the screen and the status of the figures themselves. If in actual space the bodies used by the project were particular beings with individual names and lives, their 'relocation' to the other, virtual side of the screen interface is understood to have transformed them into new and exemplary forms of being, normative entities that can stand for all other bodies. One of the historical functions of the deployment of Adam and Eve as anatomical figures is to place them as temporal norms, original, perfect bodies from which all other bodies depart as variation and degradation.

Moreover, if in actual space the bodies imaged were corpses, their digital icons are miraculously alive. It is clear that for the project the figures can lay claim to a form of vitality. At the most practical level this claim is unsurprising. After all, the point of the project is to create figures which act *as if* they are alive, to use the particular simulational qualities of virtual space to model living processes. Great energy is currently being put into the animation of the figures, or more accurately in developing 'active' kinds of animatronic capacities to augment the current repertoire of 'passive' animation techniques, which are concerned with traversing the data or dissecting it as a modelled object. Active animation involves the development of techniques for simulating joint kinematics, muscle flexion, stress dynamics and so forth under different conditions which can be set by the operator. Other kinds of animation focus on the simulation of physiological processes, blood circulation, respiration, pulsation and tremor. Haptic animation is another high priority. Surgical simulators provide a sense of touch, grasp, tissue density and differential resistance feedback to the hand of the proto-surgeon. All of these qualities are intended to enhance the usefulness of the VHP data as surrogate for living bodies in the context of surgical and clinical training, ergonomic design and anatomical modelling.

Dr Spitzer, the Director of the Centre for Human Simulation, where the original processing of the bodies was carried out, explains the name of the centre in the following terms:

SPITZER: What we want to happen here, we want this data stuff to react, to act human.
INTERVIEWER: You mean alive? Act alive?
SPITZER: Yeah sure. [*Referring to surgical simulation using the VHP data*] when you [*enter a patient's aorta with a needle*] as soon as your needle gets up near his aorta you can feel his aorta pulsing. ... So you think he's alive. Now of course [*the VHP is*] not alive ... [*but*] now for all practical purposes, for that little tiny application, I don't think I have to do any more to convince you that he is human.[8]

Spitzer clearly equates the force of animation with the force of life. In posing the figures as vital, able to support and demonstrate physiological processes and motility, the VHP partakes of a long tradition of anatomical iconography. Vesalius' osteological and myological figures in the *Fabrica*[9] are depicted in moments of dramatic action – walking, bending and posing – despite varying degrees of flaying. Subsequent anatomical texts have habitually placed their anatomical figures in various kinds of motion.[10] This tradition arises in part as a way to demonstrate teleological relations between anatomical structure and physiological function, 'sustaining the idea of exemplary action as the final expression of normative structure' (Harcourt 1987: 47). It is also thrown up to deal with the central paradox of anatomical knowledge, a science of life that owes a problematic debt to death. Anatomical practice produces knowledge of living bodies through the analysis of dead bodies, generating its models of living tissue and vital physiology from the study of the corpse. This dependence on the corpse is profoundly disavowed by the biomedical imaginary, at the same time as it exercises far-reaching effects on the medical idea of life. The spectacular animation of its anatomical figures is one of a complex range of strategies which seek to minimise the difference between living and dead bodies, and to efface the analytic violence involved in all modes of anatomical preparation. As Harcourt writes in his study of representational practice in the *Fabrica*, the use of animate anatomical figures

> provides a way of bracketing in representation the simple fact that anatomical knowledge is constituted through the violation and destruction of its proper objects in practice. The idealised, classical form of the figures, the fact that they do not read as actual cadavers, sets up a foil within the structure of the illustrations that mitigates the deadening, objectifying force of the accompanying narrative.
>
> (Harcourt 1987: 34–5)

In the inanimate space of the book-based atlas the animation of its figures can only be connoted, referring to the body's motility in the world of everyday space. The computer graphics which produce the VHP figures are themselves animate, able to model and demonstrate movement within its own terms. To that extent it provides the most recent means of posing the force of life, the motivation of matter, through the animation of traces or the automation of models.

Here I am referring to the extent to which scientific ideas of life depend upon and are conditioned by the media of demonstration, the technology which presents itself to model life. As Foucault (1972) suggests, life is not a transcendental quality but a specific historical formulation with a specifiable archaeology, the posited object of the biological sciences, an abstraction and entity to be explained and demonstrated rather than assumed. Life as scientific object is the force which animates living bodies, an elusive force which exceeds its location in any particular body. Like the forces of magnetism, gravity and entropy found in physics, it can be conceptually abstracted from any particular body which it might animate, and hence can be analysed, quantified and controlled irrespective of these bodies. Foucault (1972) locates its emergence in the shift from eighteenth-century natural history to nineteenth-century biology, but Canguilhem (1992, 1994) locates the possibility for conceiving life as abstractable force in the earlier development of automata. He suggests that it only becomes possible to conceptualise life as specifiable force (rather than transcendental and unquantifiable essence, as it was understood in the philosophy of Vitalism) when it is possible to build models which demonstrate it.

> For a long time, kinematic mechanisms were powered by humans or animals. During this stage, it was an obvious tautology to compare the movement of bodies to the movement of a machine, when the machine itself depended on humans or animals to run it. Consequently it has been shown that mechanistic theory [of biology] depended, historically, on the assumption that it is possible to construct an automaton, meaning a mechanism that is miraculous in and of itself, and does not rely on human or animal muscle power.
>
> (Canguilhem 1992: 47)

The production of automata itself produces an idea of life which is automatic, the mechanistic philosophy of the organism which is closely associated with the Cartesian body and its variations, and which poses life as an external force of animation for the otherwise inert organic mechanism.

The implication of the model in the configuration of what counts as life is demonstrated by Cartwright (1995) in her study of the development of physiological cinema in the late nineteenth century, and by Petchesky (1987) in her work on foetal visualisation and reproductive technologies. If the production of automata contributes to the abstraction of life, the creation of animate media – cinema, video and the computer – produces a space for the motivation and demonstration of traces where the invisible force of life can be visualised, and hence conceived. At the same time the space of demonstration and the space of life are not clearly separable, so that media and life are relational domains. This inseparability is evident in Petchesky's (1987) work, which delineates ways that the development of medical visualising technologies like ultrasound and the endoscope have, over the last twenty years, dramatically altered the time of viability of the foetus, shifting it to earlier and earlier points in gestation. The ultrasound video, the most commonly used form of foetal imaging, is not a transparent window into the maternal body. It is rather a technology which crops, frames and animates its objects, so that the motility and coherence of the foetus conceived as a separate being from that of the mother is an artefact of the medium, a motility, individuality and purposiveness produced rather than recorded. The ultrasound video's animating capacities enable the investment of foetal images with new forms of autonomy, new claims to viability above and beyond the life conferred by the pregnancy. Such claims are, in turn, translated into interventions and technologies for the preservation of foetal life at earlier and earlier stages.

Similarly Cartwright's (1995) and Marchessault's (1996) works investigate the extent to which the animating capacities of early cinema and pre-cinematic technologies shift the ground of life and reconfigure it in their own image. Both discuss Marey's famous studies of bodies in motion, using the pre-cinematic chronograph. The chronograph produced strips of discrete images that displayed the moving body captured in small increments of time, frame by frame. These sequences were not restricted to the demonstration of motion but were used as the basis for the distillation of the laws of movement, laws hidden until the analytic vision of the chronograph and the cinematograph made them visible. As Marchessault (1996) points out, these laws were then applied to the organisation of actual bodies, used to create physical education regimes which sought to make bodies conform to the economy of the film strip. Here the body's motility and temporal existence are reconceptualised according to the new ability of cinema to display the body's movement as a series of minute, sequential movements over microseconds. Just as film allows the image of bodily movement to be broken down into discreet sequences, so the actual body can be organised to move through these sequences in a disciplined fashion. The physiological body thus becomes a 'filmic' body, performing as if it were animated cinematically. Conversely the efficacy of cinema in making visible the processes of life collapsed the idea of life into cinema, so that the filmic apparatus was taken to be an instance of the operation of 'life itself', an image device which embodied the biological processes that it displayed (Cartwright 1995). Through its capacity to transcribe corporeal traces as 'vital script' or 'ontic imprint' (Lippit 1998) and to reprise their vitality through animation, cinema was life insofar as it brought to life that which it depicted.

CYBERNATURAL LIFE

These histories convey the extent to which the space of biomedical demonstration, the specific technical form of the medium and the modes of animation that it allows are collapsed into the theorisation and prostheticisation of life, the ways it is instrumentalised. In the case of the VHP, the attribution of vitality to the figures takes a specific energy from the elaboration of virtual space as vital space currently underway in multiple sites of technoculture and the cybersciences. Terming the figures Adam and Eve locates the space of the computer screen, a space projected out of the play of data, as a vital 'world', a new Eden for the technical production of artificial life. This is, of course, not the first instance of the projection of virtual space as a new Eden – the Apple Macintosh Computer icon, the trademark nibbled apple, clearly works off the same equation.

In many ways the material organisation of the computer can be seen to suggest utopia. The apparent self-enclosure of computed space, its ability to model and manipulate complex spatial geometries, to make and unmake visual objects, to set out complexity within strictly confined parameters, all of these qualities lend themselves to the conceptualisation of the screen as the outer surface of a microworld, to borrow Paul Edwards' term.

> Computer simulations are ... by nature partial, internally consistent but externally incomplete; this is the significance of the term 'microworld'. Every microworld has a unique ontological and epistemological structure, simpler than that of the world it represents. Computer programs are thus intellectually useful and emotionally appealing for the same reason: they create worlds without irrelevant or unwanted complexity.
>
> (Edwards 1990: 109)

The space of the screen is in this sense a space of autonomous production, a medium in which worlds can be invented, modified and destroyed. To pose computed space as a prelapsarian Eden is to specify such a world, a biographic space where forms of post-natural vivacity can be authorised.

This conceit is not only a function of the visually imaginative qualities of virtual space, but also arises out of a more specific conflation of living process with data process, of life with information.

Here I want to borrow Sean Cubitt's (1996) term 'the cybernatural' to designate this convergence between the force of life and the force of information. The cybernatural is, in his typology, one form of post-natural life, a rubric for a number of ways of thinking about 'what comes after the natural'. The natural is acknowledged here as the most problematic of categories, a limit which is constantly placed and replaced, trying to secure the distinction between what is considered the organic, the given, the locus of animate but unconscious matter, and what counts as the human, the cultural, artifice. The content of the natural is understood to be always tendentious, constantly pressed into service to some end as a self-reproducing limit, resource or truth. 'The natural is something which has constantly to be posited: either as the source or goal of the human, or as the other of the human which serves to define the human as species' (Cubitt 1996: 238). At the same time the category of the natural is, according to Cubitt, increasingly open to speculation about what he terms the 'postnatural', that is to speculation about the possibility of forms of vitality which do not find their support in the organic processes of matter which is understood to be the domain of the natural, but rather in the arena of the artificial. The cybernatural designates any practice which uses the space of the virtual screen as a space of 'second nature' through a conflation of information with vitality.

Given the persistent traffic between ideas of the organic and the artificial within the rhetoric of 'nature', and the implication of life with automata and animation discussed earlier, the cybernatural seems continuous with earlier configurations of life, rather than marking the rupture suggested by Cubitt. Canguilhem's (1992) history of mechanistic theories of the organism and ideas of life suggest that there is no positing of the natural within the history of European science that is not always/already technologically enframed. In Haraway's words, the history of nature is one of 'relentless artifactualism' (Haraway 1991: 295). Like my formulation of the idea of the posthuman, I would argue that the post-natural is a permanent implication and possibility sequestered within the operation of the term 'natural', a kind of permanent tension which is perhaps more evident at some techno-historical moments than others. Given this analysis, the cybernatural marks not a decisive movement beyond earlier ideas of the natural but rather the most recent rereading of the natural. It tries to make the natural account for shifts in the forms of commensurability between human and nonhuman technics, the new forms of exchange made possible by economies of code.

The cybernatural is generated out of the history of the interpenetration of living organisms with information systems, where the terms 'life' and 'information' act as synonyms. This history involves the exchange of the idea of information between the 'cybersciences' – Keller's (1995) term for the constellation of technoscientific disciplines like cybernetics, systems analysis, communications engineering, operations research and computer science, which have their beginnings in the information theory of the 1940s – and molecular biology research. In its first articulation in communications engineering, the concept of information was extremely specific – a purely statistical, content-independent technique for coding and transmitting messages, a means of securing the integrity of the signal against the depredations of noise (Oyama 1985). Once set in train, information proved a viral term, mutating and transmitting itself with protean energy into various technological practices. Most pertinently, it has been taken up in genetic research to analyse the nature of the gene, and in the cybersciences to specify the organisation of technical systems. The term 'information' is made to work in these arenas in quite different ways at the level of technical performance. As Oyama (1985) discusses in great detail, the idea of programming information in computation, and of morphogenic information in the organism, have few points of technical correspondence:

> [A computer program] has features like look up tables and control lists for deciding outcomes *just because a computer lacks the biological structure and dynamics of an organism.* Hofstadter observes that the many rules and strict formalisms employed in computer programming are ways of telling an inflexible machine how to be flexible. Biological processes, however, frequently are flexible, and are the very phenomena the programmers are attempting to imitate. A program for simulating ontogeny, or even a small part of it, would have to include not only genomic structure, but descriptions of all the conditions, parameters and interactions, internal and external to the organism, that constitute the developmental system in transition. In the biological system, the 'decision' or 'rule' needn't be programmed symbolically. ... The events themselves are 'controlled' by reciprocal selectivity ... codetermined by the system and conditions.
>
> (Oyama 1985: 61–2)

If the itinerary of the term 'information' varies in the domains of computer programming and developmental biology in a strict sense, the protean power of the term has nevertheless set up relations of commensurability and reciprocity between the two domains. Since the establishment of the respective fields, molecular and developmental biologists and computer scientists have drawn extensively upon each other for formulations about system, organism, code, feedback and transmission (Keller 1995). This reciprocity is not merely conceptual; rather the circulation of information between these different domains has in turn generated new assemblages of human and machinic organs. The pharmacology of much HIV prophylaxis, for example, depends on the engineering of drugs which interfere with the replication or transcription of viral genetic information. Biofeedback loops enable the engineering of advanced biotechnical systems like heart pacemakers, or pilot's associate technology that monitors and evaluates the brain waves, eye movements and galvanic skin response of pilots as part of its management of flight trajectories (Gray 1995a).

At its highest level of generality the engineering of technological and organic matter as information systems contributes to the projection of computed space as itself a world, a second (post-)nature which encloses a specific and relatively autonomous form of generative time and space. This is made clear in Helmreich's ethnography of Artificial Life laboratories. A-Life is a speculative scientific project that involves the writing of computer programs which attempt to mimic the emergence and evolution of living systems. A-Life scientists create what they consider to be data organisms, by writing particular kinds of algorithms, or bit strings, which combine and develop in autonomous ways out of basic programming rules. Scientists can begin with one or two 'ancestor' strings, and end up with thousands of bit string mutations, which develop, combine and recombine, delete or replicate in unpredictable ways. A-Life is intended as a methodology for modelling living systems from bottom-up synthetic methods rather than top-down, analytic methods.

> Instead of studying a rain forest top down, starting from the forest as a whole and dividing it into species, we unleash within the computer a population of interacting virtual 'animals' and 'plants' and attempt to generate from their interactions whatever systematic properties we ascribe to the ecosystem as a whole.
>
> (De Landa 1997: 18)

As Helmreich points out, the scientists involved in A-Life research understand their simulations as worlds because they understand the world to be fundamentally informational. According to these scientists,

> worlds are fundamentally made of languages like mathematics, and because computers are in their essence formal languages, computers are environments that can express artificial worlds. The formal, linguistic (and I take 'language' to include mathematics) definition of a universe allows people to take literally the idea that a universe is an abstract entity that might be implemented in different media. If the universe is simply the manipulation of information, then universes can easily be transubstantiated (or 'ported') from one system of hardware to another.
>
> (Helmreich 1994: 6)

Hence the life which A-Life works to replicate is, according to this logic, 'implementation independent – autonomous and divisible from material substrate', a quality which can be readily abstracted from the 'biochemical wetware of protein life' (Jonson 1999: 48), and implemented in another medium, the silicon substrate of the computer. Moreover, vitality has been conferred on information in a much more specific conceit, initially popularised by technoscientific philosophers like Norbert Weiner and Erwin Schrödinger, who vitalised information by casting it as the opposite of entropy, a life force which transmits and produces order in the face of the tendency of matter towards inertia and death. Weiner's (1948) text *Cybernetics* describes a history of the body from 'an economy of energy to an economy rooted in the accurate reproduction of a signal' (Tomas 1995: 23), which proposed the force of life itself as informatic.

> As entropy increases, the universe, and all closed systems in the universe, tend naturally to deteriorate and lose their distinctiveness, to move from the least to the most probably state. ... [But] there are local enclaves whose direction seems opposed to that of the universe at large and in which there is a limited and temporary tendency for organisation to increase. Life finds its home in some of these enclaves.
>
> (Weiner 1968: 15)

Weiner explicitly postulates organisms as themselves messages; 'Organism is opposed to chaos, to disintegration, to death, as message is to noise' (Weiner 1968: 85). Similarly for Schrödinger (1944) information is organised animation, the force which maintains the organism's formal activity against the counterforces of inertia and the inexorable action of the second law. The characteristic which allows organisms to maintain animation and organisation in the face of entropy can be found, according to Schrödinger, in the code script within the genetic molecule, the information in the gene, which determines the organism's ontogenic development, containing as it does 'the entire pattern of the individual's future development and its functioning in the mature state', its four-dimensional pattern in both time and space (Schrödinger 1944: 22–3). This code script is a device which allows the organism to produce 'negative entropy' by absorbing orderliness from its environment and maintaining its stability in the face of the forces of 'maximum entropy, which is death' (Schrödinger 1944: 76).

The cybernatural plays out these exchanges between information and vitality, nowhere more elaborately than the field of A-Life research. A-Life gives explicit articulation to the ways in which life and the space of its demonstration are reciprocally projected and mutually constituted through the working of information. While the modest claim for A-Life is that it 'models' living systems, its practitioners and supporters habitually substitute 'create' for 'model', on the grounds that organic life is itself fundamentally informational. Christopher Langton, one of the first architects of A-Life research, explicitly defines the difference between A-Life and biology as that between creation and analytic destruction. 'Artificial Life is simply the synthetic approach to biology: rather than take living things apart, Artificial Life attempts to put living things together' (Langton 1996: 40). Like that other space of bioproduction the laboratory, computers are, for A-Life research, a technical system through which life can be simultaneously produced and studied, a tool devoted to 'the incubation of information structures' the support for 'informational universes within which dynamic populations of informational "molecules" engage in informational biochemistry' (Langton 1996: 50–1).

A-Life explicitly theorises digital data as a synthetic genetic code, and its data entities as organisms *in silico*. The form of both organic life and data entities is taken to be a mathematical program, and the computer provides the complex informational environment to support the essential (for A-Lifers at least) features of life – autonomous self-replication, mutation and evolution. Ray makes the seamless interchangeability of genetic with computational code quite explicit.

> The organic genetic language is written with an alphabet consisting of four different nucleotides. Groups of three nucleotides form sixty-four 'words' (codons), which are translated into twenty amino acids by the molecular machinery of the cell. The machine language is written with sequences of two voltages (bits) which we conceptually represent as ones and zeros. The number of bits that form a word (machine instruction) varies.
>
> (Thomas Ray, cited in Helmreich 1998: 229)

So if information is vital, the data space of the computer is pure vitality, a medium or environment which is only information, freed from the fatal depredations of entropy found in organic space. It is, literally, a matrix, a techno-maternal space for the production of new or altered forms of vivacity, and for the immortal preservation of form in the eternal space of data. Unsurprisingly, A-Life research is saturated with Genesis rhetoric, a rhetoric which makes explicit ways in which this particular formulation of 'information/entropy as life/death has been energised by a much older dichotomy between form and matter which pervades the narrative of Genesis.[11] 'Information' as that which 'in-forms', which donates form, inherits a God-like power to authorise form. Oyama begins her interrogation of the ontogeny of information by commenting on its theological underpinnings.

> In the Western religious tradition, God created the world by bringing order to chaos. By imposing form on inchoate matter. He acted according to a convention that was very old indeed, one that separated form from matter and considered true essence to reside in the former. ... In an increasingly technological, computerised world, information is a prime commodity, and when it is used in biological theorising it is granted a

kind of atomistic autonomy as it moves from place to place, is gathered, stored, imprinted and translated. It has a history only insofar as it is accumulated or transferred. Information, the modern source of form, is seen to reside in molecules, cells, tissues, 'the environment', often latent but causally potent, allowing these entities to recognise, select and instruct each other, to construct each other and themselves.

(Oyama 1985: 1–2)

The viability of A-Life claims depend upon the digital trace, the bit strings which evolve, reproduce and hybridise, being read as (in)form(ation) whose material substrate is simultaneously incidental – it could be anything (silicon, 'wetware', etc.), because information is fundamentally portable – and indispensable, in that its materiality is indistinguishable from its form in the single register of binary code. As Hayles (1996) points out, the 'bodies of information' which are attributed to bit strings in an A-Life programmer's discourse, 'are not, as the expression might be taken to imply, phenotypic expressions of information codes. Rather, the creatures are their codes. For them, genotype and phenotype amount to the same thing; the organism is the code and the code is the organism' (Hayles 1996: 151). The 'body' of the data organism is not programmable matter, a morphology instructed by code. Instead it is isomorphic with the algorithm. This collapsing of genotype and phenotype into the 'body' of the algorithm cannot lay claim to be a vital index of life elsewhere, in the sense that Marey's chronograph discussed above could, a reprise of pre-existing, particular vitality. Rather, it lays claim to being a vital algorithm, a kind of trace which plays out within itself the emergent logic of life, giving it a speculative expression. A-Life is a modelling of life which takes on a life of its own, because life, always a partial, mutable and tendentious quality, has been refigured in the image of its informatic model. On this code-centric model it is information which motivates and animates incidental kinds of matter. As Vasseleu puts it, 'Life, as we know it today, is genetically recombinant, translatable, reiterable, sequential ... susceptible to repetition, exploitation and reinscription' (Vasseleu 1996: 112).

Programmers, the authors of form in the A-Life programs, understand their practice in terms which are simultaneously genetic and Genesis-like. Helmreich reports:

When I asked one researcher how he felt when he built simulations, he replied 'I feel like God. In fact, I am God to the universes I create. I am outside of the time/space in which those entities are embedded. I sustain their physics (through the use of the computer). I look like I am omniscient to the entities within that physics, and so on.'

(Helmreich 1994: 4)

The Genesis rhetoric in A-Life practice is free from qualification because the 'creation' involved is conceptualised as an act of pure information. Programmers 'authorise' their creatures in acts of invention which are simultaneously fictional and technical, drawing on writing systems which are themselves driven by technologies of self-reproduction and mutation, the technologies of the 'program'. As fictional acts they create narratives which are also digital entities, and as technical acts they create new operational possibilities for the narrative itself. If the space-time of the screen is itself posited as a world *in potentia*, then these forms of invention can be made to appear as theological acts of creation, a production *ex nihilo*, which does not so much give form to matter as put form into play and stand as the point of origin of time in the world of the program.

This genetic hubris, the claim to authorise life, evidently relies on a constitutive exclusion of the particularity of matter from a consideration of the emergence of living entities. While A-Life science claims to model the contingency and autonomy of life forces, its 'hardcore code-centrism', to use Jonson's (1999) phrase, involves the bracketing out of another crucial kind of complexity, the interramification and reciprocal dependence of genetic information with the material unfolding of the organism.

> To reduce life to code or program as origin or essence free of material constraints is incoherent ... radical reductionism fails to capture the complicitous and mutually contingent play of information and matter that produces vitality as it fundamentally transforms these terms and the dichotomous logic with which they are bound up.
>
> (Jonson 1999: 49–50)

Jonson (1999) Keller (1995) and Oyama (1985), among others, demonstrate the final untenableness of claiming the gene as a point of origin and first cause for morphogenesis, where 'encoded information is always sequestered from and ontologically and temporally prior to the material development which it causes' (Jonson 1999: 50). Rather extensive amounts of genetic research suggest the relationship between gene and morphogenesis as one of contingency, mutual modification and reciprocity, in which neither code nor soma can be granted a prior or causal position. In this regard the code-centrism of A-Life research and the related forms of genetic research play out IatroGenic desire, a desire for programmable matter, which unfolds in obedience to a definable set of genetic instructions. Such matter would deliver a mastery over the genesis and being of the organism, its stabilisation within regimes of technical and biological rationality. Programmable matter would ensure that the organism's degrees of contingency and mutability are already circumscribed at their moment of technogenesis. As we will see, the Visible Human Project, and its enactment of codified vitality, mobilises a similar repertoire of commitments and exclusions as those found in A-Life.

THE RESURRECTION OF THE BODY

The deployment of Genesis rhetoric in the VHP also summons up a cybernatural space, an informational Eden sequestered in the space of the computer. The iconographic 'preservation' of the figures and their complex animation cannot lay the same strong claim to cybernatural life as that found in A-Life discourse. A-Life's claim derives not only from its location in informational space but also from the capacity of algorithms to model agency and autonomy. The vivacity suggested by the names Adam and Eve in the VHP is of a more limited and passive kind: the compliant vivacity of the test subject, whose physiology and anatomy is objectified and stabilised within clinical or laboratory regimes.

The resort to Genesis iconography in the Visible Human Project is modified in particular ways, a rewriting motivated by the fact that it must negotiate a much more complex creation story than that found in A-Life. Adam and Eve in the VHP are not posed as creations *ex nihilo*. Rather they appear as recreations, reanimations which have given new life to the lifeless, to figures in whom life has been extinguished. The journalistic and scientific rhetoric generated around the VHP is full of references to and metaphors for the resurrection of the figures. This rhetoric clearly indicates a desire that the figures should be understood as reanimations of vitality, reborn into some new form of life on the far side of the screen.[12] Headlines proclaimed 'Executed killer enjoys immortality on CD-Rom', 'Executed monster will live forever', and 'Visible Woman, Man are valuable slices of life'.[13] One article observed, 'it has been possible to access Jernigan's immortalised body through the ... Web since November 1994'.[14] Another referred to the Visible Woman's 'reincarnation in cyberspace'.[15] One hypertext article on the Visible Woman concludes, 'The anonymous woman who donated her body to science to be scanned, sliced and digitised, will be reborn many times in the name of science' (Wirthlin 1996). *Science*'s article about the project states:

> Meet the Visible Human. In the real world he was a thirty-nine year old prisoner who was executed by lethal injection in Texas. But now in the virtual world he has been resurrected ... to star in the National Library of Medicine's gruesomely fascinating effort to create a comprehensive digital atlas of the human body.
>
> (Waldorp 1995b: 1358)

The detour of Genesis rhetoric through a narrative which more properly belongs to the New Testament, or, on an abject reading, to Shelley's *Frankenstein*, is motivated by the necessity for the VHP to simultaneously acknowledge and deny a complex debt to the bodies used in the imaging process. Anatomy, as I argued earlier, must always negotiate the problem of its debt to death and the corpse, a negotiation assisted by the Cartesian dualism which is foundational to the practice of scientific anatomy. If for medicine in general and anatomy in particular complex subjectivity can be partitioned into immaterial mind and mechanical body, the significance of death for an understanding of the body can be minimised. If the body is mechanism, death occurs when the mechanism fails or runs down. It is within the terms of this conceit that a power of animation, the technical motivation of traces, can count as a power of reanimation, the 'bestowing of life upon lifeless matter', to use Shelley's famous phrase. To cast the image-figures in the VHP as kinds of resurrection is both to claim that they are forms of the original bodies and to cancel out the effects of death on those bodies.

At its strongest, the rhetoric of resurrection poses the images as *perfect* translations. They are entities created by the substitution of one content for another, without loss or residue, data substituted for flesh. Here the VHP betrays the same valuation of form over matter so clearly articulated in A-Life research. If the materiality of flesh, its density, recalcitrance, palpability and opacity were to be considered as positive values, the process of production of the VHP figures could only be evaluated as a violent procedure, one which annihilates bodily substance in favour of the production of navigable spectacle. If materiality is subordinated to form this process of production reads quite differently. On a form/matter distinction, the only significance of matter for anatomy is its yielding of form to the gaze, whereas the palpability of matter, its ability to encounter and resist the touch, is simply an obstacle to the eye, an inert and incidental biomass.

On this logic the cryosectioning and photographing of the corpse constitute techniques for the temporal and visual 'arrest' of the iconic form of the body. Frozen soon after death, the form of the body is preserved at the moment before organic decomposition sets in, and the process of cryosectioning substitutes a technical decomposition driven by the logic of photographic optics. This technical decomposition acts to effectively remove the body's iconic form from organic time, rendering it up to the camera to be 'arrested' in digital photo space, at the same time as its biomass is reduced, in a reversal of Genesis, to dust.[16] Here the VHP shares a technical redistribution of bodily temporalities with a growing number of biotechnologies, which work through the instrumentalisation, the slowing down, speeding up, arresting and reiteration of bodily cycles and capacities, the animation not of image but of flesh.[17]

Taken out of organic time, the time of death and decomposition, this iconic form is made amenable to the virtual time of computed space, the time of storage, retrieval, the morph and the series. The attraction for medicine here is the attraction of reversibility, an impossible temporality in which data objects can be destroyed and then restored, repeated and deleted, deformed and reformed, morphed from one form into another without damage, loss or labour. If actual time is the time of entropy, time as constant nonreversible loss and the tending of matter towards dissolution, then virtual time is negentropic eternity, without direction or consequence. In Sobchack's (1997) terms the morphable digital object is palindromic:

> [it] can be read forward and backward, without a change in meaning ... pure change as form ... cleansed of causation. ... Transformation and all its consequences are assimilated to the notion of eternal return ... to a totalised circularity in which everything is subsumed and becomes equal.
> (Sobchack 1997: 45)

This palindromic temporality is so attractive for the biomedical imagination because it plays out the IatroGenic desire for a body which acts as mechanical, rather than chaotic system. If the body can be posited as mechanism, this implies the possibility of its programming, the abstraction of a set of formal laws from the material substrate of the machine which can be used to specify its ordering in a stable relation of cause and effect. Mechanisms, as Serres (1982) points out, stand in a particular, reversible relationship to time, which in turn implies an at least potential mastery over the temporality of living bodies, the arresting of the movement towards death. A mechanical system, he writes,

> [is] a set which remains stable throughout variations of objects which are either in movement or relatively stationary. ... Within a set of mobile material points governed by a law – Newton's law for example – it is clear that time is fully reversible. If everything starts moving in the opposite direction, nothing significant in form or state will change ... the ordinary mechanical system depends on time but not on its direction.
> (Serres 1982: 71)

If the human body could be epistemologically stabilised as mechanical system, then medicine's therapeutic interventions could potentially return the diseased living body to original conditions, a return to equilibrium, and the problem of death could be postponed. In this sense the aesthetics of the VHP act to defend against a full recognition of the hypercomplex formation of the living organism, in the sense suggested by my earlier characterisation of iatrogenic illness. Iatrogenic illness, the flux and eddy of health and illness, the uncertainty of any medical intervention into a field of disease, and the nonreversibility of the organism's being towards death; all these and many other organic phenomena suggest the living body as hyper-complex system. Rather than conceptualising the body as an organic stability, a point of equilibrium or homeostasis, from which it departs in illness and to which it returns in health, the notion of hyper-complexity suggests the living body in a perpetual movement of non-reversible imbalance, an aleatory and uncertain flow. As such, it embodies technical interventions not simply as mechanical additions or rational subtractions but rather as another element in the complex production of uncertain outcomes.

Anatomy, by definition, works on an analytic assumption that treats the body's organisation as mechanism, insofar as it treats it as a composite in which the anatomical parts add together regularly to compose a whole. In this sense its understanding of organic system is non-dynamic, or its dynamism is always restricted to the equilibrium of mechanical dynamism, the irrelevance of time to its working. Complexity theory, on the other hand, suggests an organic order which can't be deduced from methods of decomposition, in that the working of an organic system is determined by a non-linear and non-additive involvement of each part for every other, and the system's general openness to the world, its continual reciprocity with an environment. Hence living organisms, posited along these lines are, to borrow Serres' (1982) phrase, not systems as 'standing-reserves'[18] but systems as *homeorrhetic*, a singularity which exists in the perturbation of an irreversible flow, and whose behaviour may be sensitised and potentiated by unpredictable, interactive outcomes.

The aesthetics of the VHP hold all of these possibilities at bay, it seems to me, performing a tension within the disciplines of medicine itself, between the mechanical commitments of anatomy and the more potentially complex commitments of immunology, for example.[19] At the same time its fantasies of mechanism help to shore up the category of the Human as itself stable and closed to temporality, the Human defended against the implications of death.

REPRODUCTION

If only form matters for anatomy then the bodies imaged for the VHP can be resurrected as form with only digital 'substance', entropic biomass replaced by incorruptible data. Like A-Life entities, the materiality of matter in the VHP is simultaneously collapsed into and excluded by information. The VHP data-bodies can be preserved in the pure positivity and stable life of data, a prelasparian vitality before the fall into time, matter, entropy and mortality. While the VHP evokes the utopic non-time of the time before the fall, it also inadvertently brings another Christian utopia into play: the end of time, the time of resurrection and redemption. Bynum (1995) reminds us of the centrality of this utopia for Christian theology; 'Christian preachers and theologians from Tertulliun to the seventeenth century divines asserted that God will assemble the decayed and fragmented corpses of human beings at the end of time and grant them eternal life and incorruptibility' (Bynum 1995: 239 cited in Rabinow 1996: 146–7). To resurrect is both to return to a life freed from death and to redeem the whole body from fragmentation, to repair its losses. In this sense the VHP contrives to bring together and reconcile the anti-Christian rationalism of anatomical attitudes towards the dead body with Christian eschatology, a reconciliation which, no doubt, contributes to the VHP's popular appeal.

On this immaterial reading, the VHP figures are not *ex nihilo* creations. Rather they are reconfigurations of the original organic bodies, which have been 'ported' out of one kind of space, matter and time and into another, the virtual Eden of computed space. What has been de-animated in actual space, has died or been killed, can be re-animated in the microworld of the virtual screen. Unlike the claims of the A-Life scientists, whose creations need only maintain a highly abstract relationship to the living entities that they are supposed to connote, the VHP scientists must acknowledge the anatomical veracity of their icons through the maintenance of some kind of 'true' relation with the organic bodies which they utilise as original substance. These figures cannot be Adam and Eve in the sense that they were authorised from nothing; rather the authorisation of the figures is, as I stated earlier, more like that of a faithful translation, the reconstitution of an essential content in another medium or language.

As translations the VHP would seem to simply reorder the existing bodies, creating a second order creation which is legitimated as a good copy. The IatroGenic power of the figures resides then not in their status as pure creations but rather in their power of duplication and repetition. To refer to the figures as Adam and Eve is to place them at the beginning of a generative moment, a point of ancestral origin for digital reproductions of themselves. The VHP is a valuable object for medicine precisely because it can be perfectly, endlessly copied and circulated, and so can be disseminated as a new global 'norm' shared by laboratories and clinics throughout the world. Here IatroGenesis designates the specific temporality of the invention, an inauguration of a singularity, a uniqueness, which stands at the beginning of its own repetition. Derrida writes:

> There is no natural invention – and yet invention also presupposes originality, a relation to origins, generation. ... Never does an invention appear, never does an invention take place, without an inaugural event. ... [The] event of an invention, its act of inaugural production ... must be valid for the future. ... Invention *begins* by being susceptible to repetition, exploitation, reinscription.
>
> (Derrida 1989: 28)

Invention is precisely the creation of a repeatable instance, an authorisation which is bound up with technical capacities for mechanical reproduction.

> If the act of invention can take place only once, the invented artifact must be essentially repeatable, transmissible and transposable. ... To invent is to produce iterability and the machine for reproduction and simulation, in an

indefinite number of copies, utilisable outside the place of invention, available to multiple subjects in various contexts.

(Derrida 1989: 51)

To pose the figures as Adam and Eve is to designate this unique repeatability, and to simultaneously assimilate this quality of reproducibility to a quality of reproduction. The *double entendre* involved in the term 'reproduction' is crucial to this claim, and to IatroGenic desire more generally. If within the terms of biology 'reproduction' is conventionally taken to designate processes of sexuality and sexual difference, and the uncertainty involved in the complex interplay of paternal and maternal features, it also carries a self-replicating subtext. As Keller astutely observes, ' "reproduction" is an ambiguous term when it is applied to organisms that neither make copies of themselves nor reproduce by themselves' (Keller 1992: 132), an ambiguity evident in the ease with which 'reproduction' becomes 'self-reproduction' in some biological discourse. Moreover code-centric forms of genetics shore up a tendency for reproduction to be read as a copying, insofar as all processes of inheritance and evolution are posed as either accurate or inaccurate copying of parental genes, good or bad copies. IatroGenic desire depends on a reading of reproduction as precisely the replication of life and living entities along the lines of the copy and the commodity, a desire which is inscribed in the very heart of biology. As Vasseleu articulates it:

> biology [works] as a reproductive technology, which invents a mechanism in place of an origin. The mechanism of reproductive technology, as analysed by Walter Benjamin, can be characterised by its displacement of the 'originality' of the unique, and the erasure of the difference between original and reproduction. The intention of a genetic biologism is ultimately the eradication of the unaccountable differences inherent in the reproductive capacity of living beings, and the production of an infinite reiterability of the biological sameness of living things.
>
> (Vasseleu 1996: 115)

If life is information, the computer can count as a reproductive technology, both as technique and space for the limitless replication of the VHP figures, which are both ontogeny and phylogeny, original figures in an infinite sequence. Genesis rhetoric provides a narrative which encompasses this sequentiality, accounting for and (cyber)naturalising this repetition.

Paradoxically, if the evocations of Adam and Eve connote the reproducibility of the figures, they also guard against the potentially delegitimating consequences of their serial being. To designate the figures in this way is to try to secure a stable, hierarchical relation between the original body and the copy image, to secure their identities as 'original' and 'copy' per se. Here I am drawing on Deleuze's reading of the platonic distinction between copies and simulacra, a distinction which functions to create a hierarchy of authenticity between a model and its possible representations.

> Copies are secondary possessors. They are well-founded pretenders, guaranteed by resemblance; simulacra are like false pretenders, built upon dissimilarity, implying an essential perversion or deviation. ... We are now in a better position to define the totality of the Platonic motivation; it has to do with selecting among the pretenders, distinguishing good and bad copies, or, rather, copies (always well founded) and simulacra (always engulfed in dissimilarity). It is a question of assuring the triumph of the copies over simulacra, of repressing simulacra, keeping them completely submerged.
>
> (Deleuze 1990: 256–7)

The falsity of simulacra is the falsity of a copy after the fall, a copy which loses its authentic moral relation, its obligation of truthful representation, to its original. Not simply a degraded copy, the simulacrum is a deceitful image which produces an effect of resemblance, constituted through an internalised dissimilarity to its alleged model. The simulacrum's power resides in its ability to undermine the distinction between original and copy, and hence hierarchies of resemblance as such, to engulf their logics and pervert them to its own order of simulation.

The VHP figures aspire to the status of copies in the order of the similar, secondary possessors of the qualities of their original object, which allows them to function as legitimate clinical surrogates. They risk the status of simulacra insofar as the process involved in their production is read as violence. The process of production of the VHP undermines the hierarchy involved in the logic of the copy, its obligation to double the presence of the original through resemblance. Rather than a doubling, the VHP's image homologue is produced through the simultaneous destruction of the body. Each visual file can only be produced at the moment that the flesh it photographs is dispersed and disintegrated. The copy is made to appear through and in the moment when the original disappears, so that it copies not the presence of the original but its unbecoming, an infernal copying which absorbs or erases originality as such. There is no moment of temporal or spatial stability, no point in time during which a point in the body can be simply mapped to a point in the image.

This annihilation is necessary for the production of the 'impossible perspectives'[20] which create the value of the VHP for its clinical and biomedical users, the ability of the figure to be opened out and traversed, its interior and exterior navigable as homogeneous spectacle, a vision of flesh free of inertia and opacity. This annihilation is the condition of the image's repeatability, its condition as replicable data rather than singular flesh. To pose the VHP figures as Adam and Eve is to secure the effect of a distinct moment *before* the fall into the series, iterability, a moment when the first good copies stand in a primary, stable relationship of correspondence to the bodies that they represent. Under the shelter of this rhetoric the copy series can move out from this moment with an epistemological safety, because each copy replicates a first and singular true copy, which serves as first instance and guarantee. The possibility that the figures are already fallen, fallen at the moment of their inauguration, cannot be entertained.

6
REVENANTS
Death and the digital uncanny

The Visible Human Project is, it seems, a work of desire as well as an advance in technique. In the previous chapter I conjectured that various kinds of desire could be discerned in its production and circulation. In its excessive visuality the project is symptomatic of medicine's much proclaimed scopophilia[1] and its utterly repressed pleasure in pornographic orders of representation (Kapsalis 1997). As perfectly co-operative image-objects the VHP figures make their exhaustively visualised bodies available for all forms of display and optical penetration, without recalcitrance or resistance. In its claim to invent vital icons the project is symptomatic of IatroGenic desire, a desire which circulates at large in the *fin-de-millennium* biomedical imaginary, and which is directed towards the mastery of corporeal matter and vitality along the lines of the commodity and the mechanically reproducible invention. IatroGenic desire thus works to compensate for the uncertainty of any relation between embodiment and technology, its excess to any human project. It works through a refusal to acknowledge the constitutive absence of mastery implied in this relationship, the failure of technology to fully serve human interests and human ends. The VHP is a device of IatroGenic desire precisely because it presents the spectacle of a fantasised mastery over matter, a mastery which is always deferred, promised in the future.

The resort to Genesis iconography analysed in the previous chapter is indicative of the ease with which medicine deploys mythic resources in the pursuit of its fantasies, and in the play of the biomedical imaginary more generally. I have used the term 'biomedical imaginary' at various earlier points in this text, but here I want to treat it in more detail, in order to open up a certain critical ground. The biomedical imaginary refers to the speculative, propositional fabric of medical thought, the generally disavowed dream work performed by biomedical theory and innovation. It is a kind of speculative thought which supplements the more strictly systematic, properly scientific, thought of medicine, its deductive strategies and empirical epistemologies.[2] While medicine, like all sciences, bases its claims to technical precision on a strict referentiality, a truth derived from the givenness of the object, the biomedical imaginary describes those aspects of medical ideas which derive their impetus from the fictitious, the connotative and from desire. My development of the term works off that of Le Doeuff (1989), who defines the imaginary of systemic thought (in her case philosophy) as the deployment of, and unacknowledged reliance on, culturally intelligible fantasies and mythologies within the terms of what claims to be a system of pure logic. In particular, she identifies the deployment of images and metaphoric modes of thinking as indications of excess, points at which systemic thought spills out into the speculative, the fantasmic, into desire, even while it tries to relegate such images to the status of illustration of a systematically generated idea. Moreover, the deployment of images marks points of tension, knots of paradox or ambiguity within a system which are not resolvable within its terms. Her hypothesis is that

> The interpretation of imagery in philosophical texts goes together with a search for points of tension in a work. In other words, such imagery is inseparable from the difficulties, the sensitive points of an intellectual venture ... the meaning conveyed by images works both for and against the system that deploys them. *For*, because they sustain something which the system cannot itself justify, but which is nevertheless needed for its proper working. *Against*, for the same reason ... their meaning is incompatible with the system's possibilities.
>
> (Le Doeuff 1989: 3)

On Le Doeuff's proposal, and following on from the analysis of Genesis iconography laid out in the previous chapter, the designation of the VHP figures as Adam and Eve both suggests a knot of tension in the *project* of the Visible Human Project and presents a way of marshalling and limiting the interpretation of the figures along particular lines. If images work both for and against the system in which they are located, if they present such possibilities of ambiguity, the management of their presentation, contextualisation and interpretation becomes anxious and imperative.[3] Scientific thought is peculiarly vulnerable to such ambiguity, in the sense that its epistemologies are driven by imperatives towards visualisation, towards the production of images which are held, through strict procedures of referentiality, to reveal the natural world. The act of making visible is a foundational gesture for scientific thought – picturing the microscopic and the macroscopic, picturing invisible phenomena like electronic resonance (MRI), ultrasound, infra-red – in short producing a panoply of novel images whose interpretation must be contained within scientific confines.

The management of interpretation is particularly at issue in the case of widely disseminated images like the VHP figures, circulated from the moment of their initial publicity as objects of public and popular culture. The very title of the project, a reference to the novel (and subsequent films)

The Invisible Man, immediately places it within a popular cultural order and locates it in a history of popular interpretation of medical technologies. H.G. Wells's novel, the story of a scientist who discovers rays that make him invisible, was itself a response to a dramatic shift in medical vision and imagination, the shift inaugurated by the invention of x-rays (Kevles 1997). Moreover, as I described in the first chapter, the creators of the VHP, the National Library of Medicine, ensured wide public attention to the project through well orchestrated media launches for both the Visible Man and the Visible Woman.

It is at this point of public dissemination and potentially open-ended interpretation that the attempt to manage the reading of medical iconography like the VHP becomes most imperative, and runs the greatest risks. Public dissemination is pursued as a means to secure public legitimacy. The formation of a public consensus around the desirability of biomedical and scientific research depends upon forms of nonspecialist interpretation which lend such legitimacy. Science requires a multiplicity of specialist and nonspecialist audiences, and has developed textual and visual practices which both solicit such audiences and compel assent to particular interpretations of its products and activities (Shapin and Schaffer 1985). To secure such audiences, medicine actively participates in the migration of its images out of the clinical domain and into the media of public visual culture. As the editors observe in the introduction to their anthology *The Visible Woman* (Treichler *et al.* 1998), the novel imagery of science asserts a continued presence in the mass media. In turn, the necessity for public dissemination shapes the production values and aesthetics of scientific images.

> Images derived from MRI (magnetic resonance imaging), computerised tomography (computer digitised x-ray imaging) and DNA sampling, colourised and sharpened to enhance their aesthetic appeal, regularly illustrate ... popular magazines like *Scientific American* and *Life*. ... Schematic models of HIV regularly adorn book and magazine covers.
> (Treichler *et al.* 1998: 2)

Such practices of dissemination and popularisation seek to constitute an imagined community to legitimate and validate experiments and to assimilate the nonspecialist gaze to the medical gaze, an assimilation which depends on a wide acceptance of the scientific medical interpretation of the phenomena under consideration. Nevertheless such practices of popularisation and dissemination are always open to risks of reinterpretation, a reading against, or at odds with, the canonical, scientific interpretation presented. Once medical images leave the strictly regulated contexts of the scientific media, their debt to the imaginary, the speculative, to desire, the fictive, to particular cultural genres and stock narratives, becomes less readily ignored. The intertextuality of scientific images is more evident at these points of popularisation, and this intertextuality implies that the interpretation of images by different nonscientific audiences can lead off in a number of directions and is open to various orders of appropriation.

Just as the VHP is a fantasy object for biomedicine, an aide in the pursuit of IatroGenic desire, so too is it a fantasy object for other kinds of audiences. The wide, continuing attention it has received in mass media outlets, and the sporadic objections made to the allegedly unethical nature of the project are indicative of this status as an object attracting multiple dimensions of fantasy, desire and anxiety. As I reported in the previous chapter, much of this media attention drew upon the stock narrative interpretation endorsed by the project, the Adam and Eve nomenclature ready-made to insert the project into available ideas of life and its scientific management, that is, into a generally positive evaluation of medical progress. Nevertheless I would argue that, if the images in a rational system work both for and against the functioning of the system, then the working of Genesis imagery is not only positive but also defensive. It counts as one line of force in the warring forces of signification which are at play in the VHP and other technocultural objects more generally, both proffering a particular way of reading the Project and pre-empting other available kinds of readings.

This chapter will read the VHP along another line of force, pursuing a decidedly gothic and abject aesthetics which can be readily located there, and which serves as something of a critique of the anodyne self-presentation of the project. It seems to me that the defensive action of Genesis iconography is directed against such an abject reading. My interpretation foregrounds the project's debt to death and the corpse, and the ways in which the medical idea of death and the corpse as anonymous matter encroaches on and disturbs the triumphalist humanism implied in understanding the VHP as simply another step in the march of technical progress. If IatroGenic desire claims the VHP as an instance of technically engineered vitality, an abject reading is addressed to the spectacle it provides of a new kind of cyber-death or life-in-death, and its intimation of virtual space as a haunted site.

DEATH AND THE BIOMEDICAL IMAGINATION

Part of the fascination exercised by the figures imaged for the Visible Human Project is generated by the difficulty of specifying their vital status; while the presentation of the figures as resurrections implies that the death involved in their production has been neutralised or cancelled out, the figures nevertheless seem to partake simultaneously of living and dead states. Cryogenically and photographically 'arrested' just after the moment of death is pronounced, their icons preserved in incorruptible data, the reformulated figures display both the particularity of persons and the disturbingly object-like anonymity of dissection room cadavers. They are kinds of still life, or *nature morte*, images in which life dwindles and fades, yet remains. Like all entities represented through the *nature morte* aesthetic, they are 'subjects ... arrested in the course of transmuting. They have not altogether lost their character as life forms but have begun nonetheless to take on the character of things, substances, material objects' (Young 1997: 124).

The pathos of the VHP figures is the pathos of the still life as *memento mori*, the image which both preserves the dead in the particularity of life and serves to remind the spectator of Death. For a long time in the early history of anatomy the anatomical figure retained the vestiges of the *memento mori* aesthetic; in the anatomical atlases of the seventeenth and eighteenth centuries, osteological figures are frequently depicted contemplating a skull, or in communion with an angel of death. Myological figures might be depicted seated on their own graves, or in conventional attitudes of moral anguish. (See Figure 6.1.)

If the VHP figures lack the explicit markers of this aesthetic, as animate corpses they nevertheless play out one of the more difficult paradoxes in the biomedical imaginary – the disavowed centrality of the corpse and death to the medical idea of life.

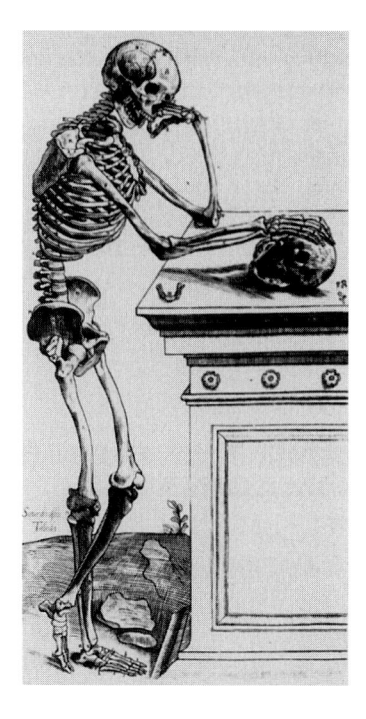

Figure 6.1 Melancholy skeleton, in *Compendiosa Totius Anatomie Delineatio* by Thomas Geminus, 1545

This disavowal asserts itself both in medicine's attempt to maintain a clean mutual exclusion between life and death, and in the constitutive failures of this distinction, its persistent collapse. As I have already discussed in Chapter 2, Foucault (1975), in the closing pages of *The Birth of the Clinic*, argues that the medical fiat to posit 'Man' as an object of technical and scientific inquiry derives from its promise to augment life against the predations of death, to exorcise death from the experience of the living. The ontological status of 'Man', his claim to self-presence and mastery of the natural, is, Foucault suggests, profoundly indebted to the ways in which medical science fends death off and casts it out, extending the vitality of the body to its furthest reach. Death is a scandal and intolerable limit to this status, demonstrating as it does the indebtedness and vulnerability of 'Man' to a contingent and wayward embodiment. Death demonstrates that flesh has its own logics, and emblematises all the ways in which embodiment exceeds or refuses to act as agent for conscious projects and desires. Medicine's unique power is its ability to technically alter the terms of this embodiment, augmenting it against the claims of death and trying to make death residual, the state to which life is opposed in a zero sum game.

This exorcism of death takes numerous forms. Within various medical discourses the deferral of death is understood to produce an intensification of life, while any practice which might produce morbidity is cast as concession to death. Activities like smoking are understood within medicalised risk discourse to disperse the time of life and expand that of death, extending its hold into the present time of the living. In Keane's words, behaviour deemed unhealthy is cast as a kind of 'disordered temporality', a soliciting of the time of death at the expense of life's duration (Keane 1997: 160). Medicine's task here is to accumulate life-capital *against* the claims of death, so that the extension of life is understood to marginalise death, to push back its limits through strategies of prevention (Tierney 1997).

Life is further accumulated against death through the medicalisation of death, the ordering of its trajectory through various forms of clinical intervention. Contemporary medicine focuses much of its technical skill on the seriously ill patient, using an array of vigorous interventions and procedures to control or manage acute conditions. If death cannot be held off indefinitely, it can at least be transformed from an unwinnable battle into a series of smaller victories, the winning back of the patient from the threshold of death through technique (Tierney 1997). For the seriously ill patient in the intensive care ward, it is only after the exhaustion of every available intervention that a patient will be considered to be dying. As Muller and Koenig's ethnography of intensive care practice indicates:

> Patients were generally considered by residents to be dying when the residents determined that there was nothing they could do to reverse the

> course of the disease and the patient would not recover no matter what they did. ... Dying is defined [as] ... treatment failure.
>
> (Muller and Koenig 1988: 369)

This technical ordering ensures that the time of dying is still secured *against* death, retained until the last moment on the vivid side of the life/death distinction. In a sense, the terminally ill patient is only conceded to be dying when they are actually pronounced dead. Moreover, each death is assimilated to the status of pathology, a disease with causes, and hence a potentially treatable condition. The causes of death are probed and diagnosed to assist in the possible prevention of, if not death as such, at least this particular *kind* of death.

> All deaths have causes ... each particular death has its particular cause. Corpses are cut open, explored, scanned, tested, until *the cause* is found: a blood clot, a kidney failure, haemorrhage, heart arrest, lung collapse. We do not hear of people dying of mortality ... they die because there was *an individual cause.*
>
> (Bauman 1992: 138)

Within the terms of this relegation of death, medicine's immense effort to exclude it from the time of life, the corpse belongs to the same margin. The corpse stands for the failure of the medical armamentarium to stabilise the body on the side of life and presence. It bears witness to the limits of medical control over natural process, and signifies the end of the time of life and the beginning of the time of death. Located within this mutually exclusive dichotomy the corpse reads as the negative of life, its opposite. This logic of opposition and mutual exclusion is played out in the legislative 'moment of death', the establishment of a clearly delineated point in time which divides the living body as subject from the dead body as object, person from corpse.

At the same time the very existence of the VHP, and the history of anatomy from which it emerges, testifies to the presence of the corpse within, rather than outside, the medical idea of life. As I argued in Chapter 3, the VHP recapitulates a whole history of anatomy, which is itself a history of the centrality of the corpse in the production of biovalue. The practice of dissecting corpses in order to produce anatomical knowledge marks the emergence of the first properly scientific bioeconomy, an economy propelled by the differential value between the corpse at the social and biological margins of life and the knowledge it generates, knowledge which can be used to enhance the well being of the living. The anatomical dissection marks the point at which *medical science itself develops out of a productive encounter with death*, the mining of death to increase the value and productivity of life, its technical augmentation.

In contemporary medicine the VHP is only one of a range of new biotechnologies which draw upon the bodies of the just-dead, the nearly-dead or the not-quite-alive to produce various kinds of biovalue. At one margin the viability of foetal life is constantly brought forward, located at earlier and earlier points in gestation through the use of reproductive technology (Hartouni 1997) and non-viable foetal tissue is utilised for the production of cell lines, vaccine development, tissue transplantation and Human Genome research (Casper 1995). At the other margin the corpse has become the basis for an array of vitalising procedures and researches. An ever-growing array of biotechnologies enable the vitalisation of the legally dead body so that the cadaver can be utilised as a donor for organs, corneas, connective tissue, bone, heart valves, cells and skin (Hogle 1995). The determination in the late 1960s of the legal definition of death – as 'brain death', the absence of signs of neural activity, rather than the earlier and more traditional signs of absence, the still heart and the absence of breath – was designed specifically in relation to new technologies which could maintain vital functions indefinitely after brain activity ceased. This shift of demarcation point allowed the *process* of death, the fact that it is a distributed temporality rather than a single point in time, to be instrumentalised in a number of productive ways. These new technologies have produced the donor cadaver, which can be connected up to a complex system of ventilators, intravenous fluid pumps, biosensors and thermosensor warmers. These support systems enable the temporary maintenance of vital function, so that the legally dead body can act as source of organs and tissue for transplants, or for pharmaceutical or medical research. Such a body is 'animate but not legally "alive", organic but chemically preserved, a mechanical and chemical object which breathes' (Hogle 1995: 204). In the case of many intensive care deaths, the distinction between living subject and useful cadaver is not perceptible apart from technical systems which indicate brain function. Like the VHP figures, donor cadavers partake of both living and dead states, and the placement of the medical definition of death at the point of brain death enables the preservation, fragmentation and commodification of certain qualities of vivacity, making these qualities available for the use of the still living.

If death is open to vivification within medical systems, this process is assisted by *a more general reliance on the corpse as an implicit model for the living body within the biomedical imagination*. Death can be rendered more life-like because the living body within medicine is modelled upon the dead. Here I am working from an argument made by Leder (1990, 1992) regarding the consequences of medicine's continued reliance on a Cartesian concept of the body as mechanical matter, animated and organised according to machinic laws. The corpse, Leder contends, is central to the medical idea of the body because the corpse embodies the Cartesian idea of the body as animated or deanimated mechanism, a machine whose animation is provided by an essentially external and disembodied subjectivity.

Leder argues that the living body presents a certain problem for medicine, in the sense that it is difficult to contain its meaning to that of predictable matter and force. The living body is excessive, unpredictable, organised through unquantifiable forces of meaning and desire, as well as complex, nonfunctional kinds of organic drive. The psychosomatic abilities of bodies, the formation of hysteric illnesses and corporeal disturbances testify to the dynamic, self-organising complexity of these elements, their synergistic interactions which exceed the sum of their parts, and which complicate the activity and agency of bodily materiality in ways still largely unrecognised by medical models.[4] As I discussed in the previous chapter, even the simplest illness is capable of precipitating unpredictable and complex cascades and perturbations, non-linear and non-additive developments which are poorly accounted for using mechanical models. It is a recognition of this non-additive quality, in for example allergy reactions and autoimmune diseases, that has led some medical researchers to deploy open systems theory to account for the operation of the immune system.[5]

Moreover, as a libidinal entity the living body cannot be securely confined to a simple, demarcated volume in space. Its boundaries and surfaces are extendable through pleasure and pain, through identification and a capacity for complex forms of embodied relationship and assemblage with other persons and things. As Grosz (1995) puts it, the body is both extendable and receptive surface:

> As a receptive surface, the body's boundaries and zones are constituted in conjunctions and through linkages with other surfaces and planes. ... These linkages are assemblages that harness and produce the body as a surface of interchangeable and substitutable elements.
>
> (Grosz 1995: 34)

These qualities of libidinal extension, protean potentiality and openness to unpredictable transformation exist in tension with the mechanistic model of the body posited within dominant forms of medical thought, a model dedicated to quantification, predictability, functionality, the homeostatic rest point and the reversibility of conditions. As Grosz points out, it is precisely the plasticity of the body, its abilities to enter into prosthetic relations, that opens it out to the normative powers of biomedicine and leads it to act 'as if' it is simply mechanism.

> The increasing medicalisation of the body, based on processes of removal (incision, cutting, removing and reduction) or addition (inlaying, stitching, and injection) demonstrate a body pliable to power, a machinic structure in which 'components' can be altered, adjusted, removed or replaced.
>
> (Grosz 1995: 35)

At the same time its participation in meaningful inscription and intersubjective animation propels it beyond medicine's reductive movement. The libidinal, Iatrogenic body is always to some extent at odds with medicalised norms of the body as isolated, stable volume, a hierarchically ordered organism whose modes of assemblage are only ever functional and predetermined. Hence the living body is never adequately accounted for by mechanistic models of cause and effect, nor by hierarchical, functional models of organism coherence. Leder writes:

> While the body remains a living ecstasis it is never fully caught in the web of causal explanation. I may attempt to understand the movement of another by tracking internal chains of physiological events. Yet the living body is that which always projects beyond such a perspective. Its movements are responses to a perceived world and a desired future, born of meaning, not just mechanical impingements. This bodily ecstasis constitutes an absence that undermines attempts to analyse the body and to predict and control its responses.
>
> (Leder 1990: 147)

Leder argues that this tension is resolved for medicine in the utilisation of the corpse as the model for the living body. The corpse is a thing that retains the macro-anatomy of the human, while abstracting it from all the forms of relation and transformation which mark its everyday existence. Hence, it removes the body from the technosocial systems which inform its complexity and vitality, so that it can be treated as a homeostatic, closed system. Its lack of dynamism lends it to analytic forms of understanding, conceptual dissection in which the parts can be abstracted from the whole because their interaction is that of the components in a rational machine, non-cumulative and already specified. It is this positing of the body as machine which forms the basis for objective diagnosis and clinical treatment, the posing of a stable system of cause and effect, treatment and cure, which holds across all kinds of bodies in all circumstances.[6] It is also the quality which makes it understandable through dissection, an act which asserts that the whole is only the sum of its parts.

The deployment of mechanical modes of understanding corporeality has profound consequences for the understanding of life and death. As Leder contends, the idea of the body's life in mechanistic biomedicine is produced through the conceptual repression of death, the minimisation of its significance for medical understandings of the living body. Death is only a matter of technical difference in state, the difference between an animate and an inanimate mechanism, and the inanimate corpse serves, briefly, as a model of matter at rest, a moment of stability and objecthood, at least at the level of observable macro-anatomy.[7]

> The body's so-called life is modeled according to the workings of an inanimate machine. The [dead, anatomised] body can constitute the place of life only because life itself has been fundamentally conceived according to the lifeless... Without the soul's presence, the body [for Descartes] would remain an operative machine, but one devoid of any truly experiential life. Dissection of the corpse can provide a method of studying the living body only because the latter is itself a sort of animated corpse.
>
> (Leder 1990: 143)

It is this idea of mechanical life, and its minimal distinction from death, which gives the study of the corpse its centrality in the history and current practices of medicine, at the very threshold of induction into medical knowledge in the anatomy class, as the means of understanding the locus of disease through the pathological anatomy, and as the template for the operation of the x-ray and other imaging technologies that anatomise the living body. Surgery is performed by making the living body as close to a corpse as possible, subtracting through narcosis the forces which distinguish it from inanimate materiality. Moreover it is the idea of life which allows the VHP figures to be posed as reanimations, kinds of living text whose debt to the original death of the bodies imaged is incidental because it is essentially reversible. If the morphology of the original bodies must be destroyed to produce the data files, the rendering capacities of virtual space can nevertheless repair this destruction, a simple synthesis of that which has been analysed. The form of the dead body is all the more amenable to 'arrest' because it is at rest, and this inertia can be reversed through animation, which is, as I conjectured in the previous chapter, fantasised as a partial restoration of vitality. In relying on the logic of reversibility the VHP partakes of the general repression of death in the medical idea of life, providing a model of the living body which is precisely a reanimated corpse.

MEDICAL HORROR

The medical idea of life is ghosted by the corpse, and the corpse by life, despite clinical protocols which mark a clear and unambiguous distinction, a moment when 'the body-as-object is definitively broken off from the body-as-self' (Young 1997: 126). If, as Foucault suggests, medicine makes 'Man' an object in order to present him with the possibility of a technically augmented vitality, a defence against the depredations of death, it does so by constantly reworking and exploiting the limit between life and death, subject and object, living being and corpse, a reworking made possible through the very insistence upon their mutually exclusivity. Hence it is through marking the point of death as 'brain death', designating this temporal point and no other as the legislative moment of death, that donor cadavers can be transformed into resources of vitality for the living, even while the cadaver breathes, its blood circulates and its heart beats.

If, as Leder (1990) argues, death for medicine signals merely a technical difference in state, its significance for subjectivity is nevertheless constitutive, presenting the spectacle of absence and non-being which encroaches on the structure and temporality of self-presence. In Kristeva's succinct formulation

> the corpse, the most sickening of wastes, is a border that has encroached upon everything. The utmost of abjection. It is no longer I who expel, 'I' is expelled. ... It is death infecting life ... it beckons to us and ends up engulfing us.
>
> (Kristeva 1982: 3–4)

The power of medical technology and discourse to make demarcations between living and dead, subject and matter is inevitably bound up with the anxiety of subjectivity in the face of that which erases it, the corpse which signifies its final exclusion from the world, and the final failure of the world to be *for* the subject. This anxiety, Young (1997) suggests, contributes to both the investment in clear demarcations between the living and the dead and to their failure.

> Pathology's attempt to reinscribe the space of death on the body as precise, to rearticulate the time of death as instantaneous, is a response to this discomfort [with death]. These are attempts to disambiguate the ontological status of the corpse as subject-turned-object by conjuring up crisp, clear, clean boundaries in space and time.
>
> (Young 1997: 127)

The medical exploitation of the shifting and blurred limits between living and dead states becomes the locus of such anxieties, introducing the possibility of technically articulated states of being caught between presence and absence, subjectivity and objecthood. In this sense the experience of medicalised subjectivity which Foucault (1975) alludes to – 'that technical world which is the armed, positive, full form of [\'Man\'s'] finitude' – is always suspended over the abyss of the abject, threatening to fall away at each moment into objecthood, monstrosity and death. Correlatively, medicalised death always presents the possibility of posthumous reanimation, or at least a continuation of residual or monstrous subjectivity after the point of medically determined death. As Young's (1997) ethnography reveals, even within the morgue, a laboratory environment dedicated to the idea of a clear distinction between the living and the dead, the corpse continues to allude to its own personhood. It retains features of identity and presence, a continuity with life not completely severed by the declaration of its objecthood.

It is this anxiety, and the ambiguity of prosthetic death-in-life, which has been figured more and more insistently as horror in popular cinema and fiction, horror which frequently engages directly with the problem of distinguishing life from death and subjectivity from technical systems or inert matter. The precise hold that medical technology exercises upon the body provides the horrific with extended scope, presenting as it does the capacity to merge or rend bodies, to penetrate and use them in parodies of medico-technical rationality. As Boss (1986) indicates in his study of medical horror cinema, the capacity of medical technology to engage and drain off bodily vitality, to suture living bodies into its own technical systems as productive components, and to bypass corporeal identity in favour of the productivity of flesh, is a potent source of anxiety for the viewer. In films like *Coma* and *Terminal Choice*, in episodes of *The X-Files*, medical technologies take over and reorientate or extinguish the agency of patients and use them to some sinister end. Medical horror suspends its victims in a state caught between proper life and death, and between an ideal of proper bodily autonomy and absorption into the machine. The horror of such technomedical horror genres is the ease with which the humanist armamentarium of medicine can be made to serve inhuman ends, to treat the human as resource and standing reserve.

> Through the image of fully institutionalised modern medicine, hospitals, banks of life-support equipment, the inscrutable terminology, the rigid regime and hierarchy, one's own body rendered alien, regulated, labelled, categorised, rearranged, manipulated, scrutinised and dissected, we experience the powerful and pervasive idea of the subject as defenceless matter becoming integrated into a wider frame of reference in which the institutional and organisational aspects of medicine ... focus their conspiratorial attention upon it.
>
> (Boss 1986: 20)

As still lives, participants in both life and death, the Visible Human Project figures are readily located within the genre of medicalised horror. While the deployment of Genesis iconography encourages their interpretation as icons of medical progress and hyperproductive resources for the generation of medical knowledge, they also present a new iconography for terminal states and bizarre prosthetic deaths. Their vitality is produced through the most exhaustive assimilation of once living flesh into technical systems. The figures' bodies have been reinvented as virtual traces, whose dematerialisation and capacities for distribution and reproduction seem to prefigure some strange new future for the body, the final collapse of even the most residual boundaries between flesh and data. As reanimations they also seem to present a contemporary variation on that most famous horror narrative, Mary Shelley's *Frankenstein*. Both narratives produce (or simulate) life through the reanimation of the dead, 'the bestowing [of] life upon lifeless matter'. Both produce whole bodies through a reversal of the anatomical dissection of criminals; the fragments of criminal corpses are assembled into the anatomy of an invented being. Both produce copies of the human in parodic acts of Genesis, figures which present a mirror-image and double of 'Man'[8] refigured as invention rather than inventor. While, as I argued in the previous chapter, the VHP claims itself as good copy of the human, in *Frankenstein* the copy knows itself to be diabolical, a bad copy. At one point the monster says to his creator 'My form is a filthy type of yours, more horrid even for the very resemblance' (Shelley 1818: 115). In both cases the figures are self-consciously likened to Adam, but for the VHP this is an anodyne Adam, a prelapsarian figure untouched by an encounter with death or with the fall. For *Frankenstein* its Adam is monstrous, a being brought forth through the technical reversal of death into intolerable, abject life, a reanimation of corpse tissues whose simultaneous debt to death and to technogenesis is visible and irredeemable.

Read within the lineage of *Frankenstein* and against the Genesis rhetoric deployed by the project, the VHP figures are mirror images, not of an essential human content, but rather of a human whose debt to both death and technics is exposed, conferring upon them the full ambiguity of the technically animate corpse.[9] The compelling aura which surrounds the figures is not limited to this moment of ambiguity, but is also generated out of their location in virtual space. The VHP figures present a new locus for prosthetic death, the space of the virtual not as a means for the generation or invention of life but as a form of afterlife. In the previous chapter I argued that for medicine, virtual space presented possibilities for reanimation which effectively cancelled the debt to the abject. At the same time the location of reanimated corpses in virtual space suggests the possibility that virtual space is a haunted site, a supernatural medium for life *after life* which supports forms of the uncanny.

THE DIGITAL UNCANNY

For medicine, virtual space is effectively a form of clinical or surgical work space, a new form of anatomical theatre where the rational ordering and workings of the body can be demonstrated to a globalised audience. The qualities of the virtual medium make it attractive as a clinical space, facilitating the complex forms of display, and multiple points of view already described. In this sense, medicine partakes of the dominant interpretation of virtual space as simply another locus for the operation of technical rationality. Here virtual space is understood as a new manifestation of the Euclidean master-space of western work-oriented cultures, the master-space which governs communication and production. As work space the virtual realm is another productive space of manufacture, experiment and communication, a technical prosthesis which supplements the productive technoscientific space of the public world through the provision of a common, transnational work environment, a transportation space for data, and a medium for the circulation of the digitised economy (Tomas 1994). It is cast as simply another transparent medium through which communication moves with greater and greater efficiency, more signal, less noise. As clinical or bioexperimental space for medical research and practice, virtual space may preserve and resurrect the dead, but it does so through a rationalistic form of technical mastery and the relegation of the meaning of death to a technical distinction, rather than through any form of miraculous power. If virtual space is imagined as a locus of vitality, the irrationality of this proposition is readily absorbed back into the assumed rationality of the medical project to invent reproducible life, to bring the forces of life into line with the technoscientific world.

Cutting across this insistence on the rational texture of virtual space is the possibility for uncanny vitality, implied in the post-natural equation between data and life. For medicine this perverse vitality gains momentum from the repression of death in medicine's idea of the body, returning to ghost the VHP. This vitality is the energy of the uncanny which Freud (1919) describes as an energy produced in the slippage between living and dead states. The uncanny, for Freud, is that anxiety which arises from ontological uncertainty about the distinction between the animate and the inanimate, objects and persons, presence and absence, the dead and the living. It is the horror specific to the possibility that the apparently living may be things, while the thingness of the dead may be illusory, concealing the possibility of secret presence, reanimation and malicious return. Freud singles out certain exemplary instances of the uncanny. An aura of the uncanny clings to automata and to dolls, producing, 'doubts whether an apparently animate being is really alive; or conversely, whether a lifeless object might not be in fact animate' (Freud 1919: 226). Automata may attain the vitality that they mimic, and the apparently human may revert to automatic states, collapse into mechanical thingness.

The figure of the double is also uncanny for Freud, a figure which works as both 'an energetic denial of the power of death' and a harbinger of death (Freud 1919: 235). The figure of the double is a mirror of the subject and a repetition of the same, both reflection and displacement. The double repeats the life of the subject, and through doubling, life is shored up against death, 'a preservation against extinction' (Freud 1919: 235). At the same time, through the possibility of displacement, the doubled presence of the self in alienated form, the double announces the subject's death, his annihilation at the hands of his mirror image.

Above all, the uncanny is figured in the revenant, the dead who return to living space as ghosts. The ghost, Freud states, is the exemplar of the uncanny because it introduces the highest degree of ambiguity in relations between the living and the dead. It testifies to the sense that nothing is ever sufficiently dead, that death can never be sufficiently repressed, and that the absence of death is only ever provisional, an abject, malicious presence held in abeyance. The reanimate dead are the familiar become alien, the impossible presence of an absent one, the human transformed into the inhuman, yet without loss of energy. The dead return to the space of the living, to the *heimlich*, threatening that seemingly secure space with their presence and reversing the temporal ordering of life and death, the replacement of the first by the second. They haunt the living with the spectre of their own death, and with the possibility that they are already, secretly dead. The threat of the revenant, Freud speculates, is the possibility it presents that the division which keeps the dead and the living separate has dissolved. Not only may the dead return to the domain of the living, but the living may be taken off to the place of the dead. 'Our fear [of the dead] implies the old belief that the dead man becomes the enemy of his survivor and seeks to carry him off to share his new life with him' (Freud 1919: 242). In each of these ways the revenant figures the confusion of life and death, 'death within life, life in death, non-life in non-death … a bit too much death in life; a bit too much life in death, at the merging intersection' (Cixous 1976: 545).

As reanimated corpses and virtual automata, the VHP figures resonate with the energy of the uncanny, altering and throwing into question the everyday sense of the relationship between living and dead bodies and the kind of spaces that they properly inhabit. Their mode of production, their dissolution in actual space and 'reincarnation' in virtual space, seems to confirm the superstitious sense that the dead leave the living world to *go elsewhere*, that presence continues in an altered form. The immortality attributed to the VHP figures in the popular press is the disturbing immortality of the ghost, its power to continue in a liminal state of non-death and non-life. If death takes the self out of time, the revenant's existence is timeless, suspended forever in the non-time of death. 'A ghost never dies, it remains always to come and to come back' (Derrida 1994: 99). As spectacles they take on the qualities of spectres, the becoming visible of apparitions who appear to the living as 'paradoxical phenomenality, the furtive and ungraspable visibility of the invisible … the tangible intangibility of a proper body without flesh, but still the body of some*one* as some*one other*' (Derrida 1994: 7).

If I can argue that the VHP betrays an aura of the uncanny, this is also because virtual space or cyberspace readily sustains a meaning as supernatural space, a meaning which Tomas (1994) and Cubitt (1996) suggest is sequestered within its dominant meaning as rationalistic work space. Virtual space is

> the space of Western geometry: the geometry of vision, the road, the building, and the machine. On the other hand this master space is a binary construct, consisting of the everyday social or profane spaces … described

in Euclidean terms and sacred or 'liminal' spaces. As Victor Turner notes, 'for every major social formation there is a dominant mode of public liminality, the subjunctive space/time that is the counter stroke to its pragmatic indicative texture'.

(Tomas 1994: 34)

For Turner this space is cinematic, but Tomas argues that virtual space is rapidly becoming the new mode of liminal space, because of, rather than despite, its adoption as primary work space. The logic of the virtual is supplemented with a mytho-logic, a sacred space which accommodates forms of ambiguity and the abject – a space open to forms of non-being, death, non-absence and nothingness which impinges upon proper identity and life in actual space.

Cubitt (1996) takes up a similar position in his propositions concerning the post-natural. In the previous chapter I worked off his argument regarding the cybernatural, one form of the post-natural, a rubric for designating forms of life which claim to transcend or succeed the life of the organic world, the world of matter. The cybernatural is the most recent articulation of the postnatural. As Cubitt points out, the cybernatural, as a form of life which transcends the organic, finds associations with an earlier, traditional form of the post-natural. That is, the cybernatural inherits the qualities of the supernatural, the life-after-death. He specifies three primary forms of the post-natural coexisting in *fin-de-millennium* culture; the supernatural, the anti-natural and the cybernatural, each of which is implicated in the other.

> These postnatures ... recognised in their simplest forms as, respectively, the life after death, the triumph of technology over nature, and the creation of artificial life, have been and are still proposed as gateways to a new era. ... [These] postnatures do not supersede or sublate one another, but coexist in the ways we think about the digital domain. Postnature is not a unified zone, any more than nature itself. With each postnatural vision comes a revision of the natural, and with each new nature/postnature pair is associated an aesthetics and a politics. But each vision persists in the others, and the more a schema is historically embedded and the more it is regarded as outmoded and forgotten, the more it returns as repressed ... in the others. Such ... is the fate of the supernatural, whose ancestral ghosts haunt our machines.
>
> (Cubitt 1996: 238)

The computer screen forms a membrane which both separates and connects two different forms of life. In the cybernatural discourse of A-Life this space is posited as a domain for new forms of non-organic vitality, a space for the invention of informatic life. The Visible Human Project's claims to post-natural vitality depend upon this conceit. At the same time, in its uncanny aspect, the VHP also partakes of the related set of discourses which posit virtual space as a haunted space, a topos for the support of spectral forms of life after death.

These discourses are discernable across a range of practices, from the science fiction of cyberspace to the research efforts of some cybernetic engineers. Much cyberpunk fiction proposes virtual space as both an alternative world and an afterworld, a space where characters in actual space can assume temporary digital personae or become permanent 'data constructs'.[10] Constructs are the digital recordings of personal essence, generated by downloading a particular subject's traits, memory and knowledge into digital data. Once recorded they can, as Stone (1992) puts it, decouple from the original subjects, abandon their bodies in actual space and pursue their own forms of life in cyberspace, as digital ghosts which inhabit landscapes of pure data. Less explicitly fictional forms of this idea are elaborated in some areas of cybernetic and robotics research, where programmers are endeavouring to write programs that will record the psychological profile of individual subjects as a way to relocate consciousness from body to computer. Hans Moravec for example, a fellow of Carnegie Mellon University, is attempting to write such programs, working from the proposition that identity and consciousness are 'essentially a pattern or form rather than a materially embodied presence. Abstract the form into an informational pattern, and you have captured all that matters about being human' (Hayles 1996: 158).

In much of this discourse, post-natural life in virtual space is imagined in anodyne or romantic metaphysical terms, where virtual space is the longed for space of spiritual freedom from the body, a freeing into digital immortality. The virtual realm is represented as exercising a kind of pull upon subjects in social space, inviting them to merge with the world of the screen.

> At the computer interface, the spirit migrates from the body to a world of total representation. Information and images float through the Platonic mind without a grounding in bodily experience. ... The surrogate life in cyberspace makes flesh feel like a prison, a fall from grace, a sinking descent into a dark, confused reality.
>
> (Heim 1994: 75)

The impulse towards the supernatural which haunts the cybernatural is clearly locatable within the same Cartesian dualism which informs medicine's ideas of life and death. Virtual space can act as a topos for the afterlife because the body, as matter, is the place where death takes place, while subjectivity, projected as spirit, can be detached from its mortal substrate, to continue after death in an altered form. As Leder (1990) argues, Descartes's mind/body distinction is propelled by anxieties about death; by making the self or soul detachable from the body, it simultaneously detaches it from the site of mortality, the corpse and corruption. The supernatural possibilities of virtual space also play upon the power of the trace, its supplementation of and for presence. If, as Moravec, cyberneticists and A-Life scientists assert, the world is fundamentally informational, then the digital trace can transmute the world into an indestructible form and relocate it inside the computer. Just as the digital trace can act as the surrogate for living entities, and for the very essence of vitality itself, so can it store and transmit the form of entities after their materiality has been corrupted. If data's self-replicating and self-evolving capacities lend it to ideas of life and vitality, its self preserving qualities lend it to ideas of afterlife, a continuity of the form of the self after the matter which 'houses' the self decays. The longing which Heim describes is simply the old masculinist, Christian dream of transcending mortality by transcending the body. At the same time this longing is the obverse of uncanny anxieties, which turn upon the possibility of incomplete decouplings of spirit from matter.

While the longing to escape the body and merge with the data flow assumes and wishes for a clean break and clear separation, the uncanniness of the revenant or the automaton are instances where the spirit cannot leave the world of matter, or where dumb, machinic matter becomes monstrously inspired. This is the paradox of the revenant. Its posthumous embodiment, Derrida (1994) states, distinguishes it from the disincarnate spirit.

> [The spectre] assumes a body, it incarnates itself ... the spectre is a paradoxical incorporation, the becoming-body, a certain phenomenal and carnal form of the spirit. It becomes, rather, some 'thing' that remains difficult to name: neither soul nor body, and both one and the other. For it is flesh and phenomenality that give to the spirit its spectral apparition.
>
> (Derrida 1994: 6)

It is as posthumous incarnations that the VHP figures haunt virtual space, visitations from the time after death into the temporality of life. Like all ghosts the VHP figures mourn the loss of the world in the loss of their bodies, and take on a shadow embodiment in order to appear in it once again. Read in this way each appearance of the figures counts as a visitation, an apparition from the shadow land of data. The menacing non-absence of the revenant seems particularly acute in the case of Jernigan's icon, whose production took place not only through an act of medical mastery but also as an extension of juridical control over his body, death and dismemberment at the hands of the state. Here virtual space is not simply the space of afterlife but an afterlife of arrest, incarceration and punishment, extended beyond execution to perpetual anatomisation and mastery by others.[11]

If the membrane between virtual space and actual space is permeable, and if in this case it maps itself onto the limit which distinguishes the living from the dead, it is not difficult to attribute to the VHP figures the power of spectral presence, a haunting of actual space from virtual space. They enact the return of the repressed which is the *revenant*, that which comes back. As Cixous writes,

> what renders [the ghost] intolerable is not so much that it is an announcement of death nor even the proof that death exists. ... What is intolerable is that the ghost erases the limit between the two states, neither alive nor dead; passing through, the dead man returns in the manner of the Repressed. It is his coming back which makes the ghost what he is, just as it is the return of the repressed that inscribes the repression.
>
> (Cixous 1976: 543)

Each time the VHP data bodies are summoned from the data banks they seem to me to carry this sense of latent force, the desire to cross back from the space of digital afterlife to which they have been committed by that repression which marks the biomedical imagination. And after all, when a body can be rendered into data and thus cross the interface into the digital afterlife, what prevents the process from effecting some form of reversal, the digital revenant who rematerialises in real space. If as I argued in the previous chapter the bodies are treated by the project as faithful translations, a full transubstantiation of flesh into data, then the process must be reversible. Faithful translation, substitution without residue, necessarily works two ways at once, creating a frictionless passage from one text to another and back again. Hence the claim that the VHP figures are surrogate anatomies, faithful copies of the real, necessarily invokes such a passage, a point of connection between the life of flesh and the afterlife of data where each can cross over to the other.

So while the kind of cyberspace summoned up by the VHP connotes the supernatural, it is not the anodyne heaven of disembodied spirituality described by Heim (1994) earlier. Rather it refers to an afterlife of the abject, the corpse which cannot or will not relinquish vitality. In the case of the VHP figures, particularly in Jernigan's case, they seem, like those other animate corpses, vampires and zombies, to be vitalised by the will of another, actively prevented from a full death. In this sense the cyberspace of the VHP presents the spectator with those anxieties provoked by the power of biotechnology to extend the residues of vitality well beyond the desires or interests of the living subject. Donor cadavers, immortal cell lines, cryogenically preserved bodies, frozen embryos – all of these marginal kinds of engineered vitality suggest an infinite deferral of death, a passage from life direct to afterlife. For the biomedical imagination this arresting and deferral of death might count as a gain on the side of life, where the VHP stands as yet another testimony to its rational, vivifying powers. Within the IatroGenic imagination, the deferral of death and the engineering of vitality appears as another step in the gradual mastery of matter, bringing it closer to the negentropy of programmable matter, the assimilation of all materiality by the metaphysics of code. For many spectators however, the VHP opens the space of a new form of death-in-life, a new and horrifying destination for our own failing bodies, and a place from which they might return in uncanny form.

7

TECHNOGENESIS
The posthuman visible

Like all executed criminals, Jernigan's virtual imago serves as an admonitory figure, a moral exemplar warning of the fate attendant on transgression of the social contract (Hutchings 1997). This doubtless accounts for much of the popular appeal of the image. It shores up a punitive sense of justice and upholds the rule of law, while translating the scene of judgement into the digital medium and giving it global distribution on the Web. I claimed at the beginning of Chapter 3 that the VHP recapitulates the entire history of anatomy. It remediates both anatomy's history of visualisation – the transition from the earliest drawn atlases, through photography, the endoscope, the x-ray and tomography – and its production of social hierarchies of biovalue, through a morality of transgression, condemnation, sacrifice and post-humous redemption. The figure of Jernigan reminds us that this history does not belong to the brutal past of biomedicine but is with us still. The Visible Human Project indicates the latent status of all citizens, all subjects, as possible objects within the optical field of biomedicine. All subjects are potential standing reserves and sources of biovalue, raw materials for biotechnical projects. Some, like the woman who donated her body to the project, take up this potential as part of the duties of citizenship. Others, like Jernigan, are designated by regimes of crime and punishment, or related regimes of class, race, sex and colonisation, as suitable material for biotechnical transformation, fated to act as test-technologies for other, more worthy citizens. Bodily matter that is deemed socially valueless can be redeemed by working its biological qualities in the interests of those whose bodies count in the social order.

Jernigan's figure, and the figure of the nameless woman, admonish in other ways as well. The strange fate of these dead bodies acts as a warning to those who continue to maintain a set of increasingly problematic humanist distinctions, between genesis and technogenesis, between social surface and the depth of interiority, between virtual and actual, and between machinic and organic. Throughout this book I have endeavoured to use these figures as a way of demonstrating the interramification of these apparently oppositional terms, existing as they (and so many other biotechnological products) do, at the nexus where these distinctions collapse and flow into each other.

Here I simply want to recapitulate some of the ways in which the VHP figures disable, or at least trouble, these distinctions, and throw them open for heightened critical scrutiny.

As we have seen throughout the book, the Visible Human figures bear witness, first of all, to the instability of the distinction between actual and virtual space. This is perhaps their most dramatic achievement, demonstrating as they do that once living bodies and social identities can be reconfigured *in silico* and strangely reanimated on the far side of the computer screen. They provide definitive evidence that the screen interface is permeable, permitting osmotic flows from one side of its membrane to the other. No doubt the uncanniness of the figures' transformation into data and relocation to the far side of the screen will eventually be forgotten, just as the uncanniness of the photograph, that other preserver of the dead, has been largely absorbed and naturalised. There is, after all, a ready-made set of relationships between user and virtual image which can be deployed to defend against the implications of this permeability of the interface.

The discipline of anatomy, and the epistemology of science more generally, furnish these relationships. For biomedical anatomy, the screen space is another kind of anatomy theatre, the Visible Human figures are bodies to be rationally anatomised, and the user is both audience and anatomist. The entire purpose of the anatomy theatre is to secure a distinction between the audience as subjects and the criminal corpse as object of knowledge. The rational spatial order and sober ritual of the anatomical demonstration clearly situate audience and corpse on either side of a divide. On the one side it locates the living subjects to whom knowledge of the body is addressed as both theory and therapy; on the other, it situates the dead, yet exemplary body of 'Man' as object in the third person (Sawday 1995). If the anatomy theatre takes place in virtual space, the interface inserts itself as a distinction between the subject and object of knowledge, safely located on either side of the terminal screen. The user can manipulate the image with a one-way touch, the masterful touch of the cursor or virtual scalpel, while the image-object is deprived of agency or the power to return the touch. The interface acts as a prophylactic barrier which allows only unidirectional action from subject to object, while protecting against any other kind of encounter.

Nevertheless, the interface cannot assert and stabilise this divide in an absolute sense. Virtual space is, like any projective technology, also a space of projection and implication for the user. It always conveys an intimation that, like the VHP figures themselves, the user is invested and located in the space of the virtual in non-voluntary ways, not outside but inside the mechanism. In this sense the VHP figures, particularly the named figure of Jernigan, solicit identification and threaten an identity which tries to contain itself safely beyond the limits of the virtual. The VHP figures exemplify not simply scientific objects but possible fates for the self whose investments in virtual space threaten absorption. In the same gesture they demonstrate the profound redistribution of space attendant on the new media. Digital media, like all media, involve a fracturing of space, a delocalisation and relocation which changes the functioning and structure of location as such (Weber 1996). This is the power of digital media to fracture familiar spaces, to upset the sense of place as *heimliche*, as safe, stable location with a specifiable relationship to the self and the world which it sustains. It is this dislocation that gives the VHP figures their spectral presence, the sense that they are located in, and emerge from, some new and uncertain zone which impinges on quotidian space in potentially threatening ways.

The figures also admonish those who attempt to read the human body as a social surface produced by subjective depth. This is, as Grosz (1994) observes, the normative interpretation of the western, humanist body, a social surface which conceals yet signifies the workings of a psychic interiority. What is essential to human consciousness and human value takes place in the interior, concealed from and occasionally revealed by, the social surface. Within the terms of this humanism, 'we are not so much surfaces as profound depths, subjects of a hidden interiority ... bodily markings can be read as symptoms, signs, clues to unravelling a psychic set of meanings' (Grosz 1994: 138–9). This humanist commitment to the unknowable interior is necessarily destabilised by any attempt to treat the interior as itself surface. While the metaphysics of the interior may claim to take place in a psychic atopia, the unlocatable unconscious or soul, this claim to interiority has nevertheless relied upon an equation between psychic and literal interior, that which exists beneath the superficiality of the body's social and organic surface (Lippit 1996). Hence the dramatic effect of the x-ray on the apprehension of subjectivity (Cartwright 1995), when for the first time the living interior of the body could be imaged, made available as an external image, a projection of interiority.

The Visible Human figures take their place in this history by abolishing all distinction between surface and depth, demonstrating that all interior spaces are equally superficial, that all depth is only latent surface. It is this movement that lends them a certain pornographic air. Like pornography, they imply that what can be seen, what is superficial and hence visible, is all there is. While erotica may imply motivation, affect and narrative – in short, interiority – pornography is only concerned with showing sexual acts. No subjective depth is attributed to them. They simply take place. Moreover if the VHP and its cognate technologies can produce the body interior as infinitely available surface then there is no depth of embodiment which is free of technological contamination. Like all those other biotechnologies which can work the body from the molecular level upwards, the VHP and related imaging technologies can order the endosoma according to complex technosocial economies and assemblages. It is no longer only the body's surface which is socially marked and produced. The entire organ-ism is open to 'superficial' modes of inscription.

In this regard the VHP figures and their modes of display suggest a new kind of interiority for the new millennium, an interior space whose superficiality and externality are always latent and articulated through instrumentation. It is a mode of interiority from which all claims to depth or essence have been subtracted. More tellingly, it is a mode of interior space which is coextensive with virtual and computational space, an interiority calibrated as data. The VHP's iconography suggests that embodied subjectivity can no longer be considered vertically, from surface to depth, but must be considered horizontally, in terms of its technosocial assemblages. If 'Man' wishes to claim a new profundity or transcendence, it will necessarily be articulated through a transparent dependence on and complicity with the computer, rather than through an appeal to any natural capacity.

As virtual *memento mori* the VHP figures admonish those who would essentialise the qualities of life and death, or consider them to be mutually exclusive and fixed states. As I have argued throughout this book, the qualities of vivacity and its absence are unstable, and their specific content depends upon technological regimes of vivification and their particular limits. Life as abstractable force is always to some extent a product of technogenesis, and life as it functions today is the life of information. The vivification of information has changed the terms of the relationship between the living and the dead. Dead animals can be cloned and revived, dead humans can be cryogenically preserved or maintained on life support systems, tissue samples can be cultivated to produce immortal cell lines or to regenerate organs, computer code becomes artificially lively, and corpses can be reanimated as Visible Humans. Death has become increasingly uncertain, and its borders are constantly breached and reconfigured through new forms of technogenesis, new ways of producing biovalue at the margins of living process.

This uncertainty has equivocal effects for those whose bodies are caught up in biomedicine's vivifying powers. On the one hand death and bodily disorder can be managed, postponed and relegated, the condition which medicine fulfils in order to have legitimate access to patients' bodies. On the other, biomedical interventions produce their own iatrogenic effects, a technological redistribution of illness and vivacity which confronts the patient with the failure of biotechnology to simply extend their will. Biotechnical configurations work by demanding that the body conforms to their logics, logics which are frequently equivocal, painful, alien or confronting for the embodied subjects enrolled into them. They undermine any claim that consciousness transcends the body by demonstrating how easily the quotidian certainties of subjective life are destroyed in the extremes of bodily experience. Moreover the reworking of death, its increasing relativisation, promises the proliferation of biotechnically augmented states of life-in-death, the possibility of suspension in some strange state between the two. The VHP figures could be the icons for such states, patron saints for medicine's new practices of revival and regeneration, and its new modes of semi-death and posthumous life.

Finally, the VHP figures admonish those who still wish to sequester the biological and the natural from the technological, or 'Man' from any of these domains. As this book has argued, the VHP is only one of a number of dramatic biotechnical developments which demonstrate that the natural and the biological are the most negotiable of entities, constantly engaging with new arrays of technology and forging syntheses, non-additive, non-divisible mixtures of the machinic and the organic. The pluripotent qualities of biological entities lend themselves to the inventiveness of biotechnical assemblage. They can be coaxed, addressed, incited, calibrated, worked over, into myriad economies and systems that themselves depend upon earlier biotechnical economies. Genetic engineering, for example, is a prosthesis for older techniques of selective breeding, while the VHP production process works off an entire history of visual prosthetics. Both of these innovations render flesh into compromise formations with data, so that earlier modes of corporeal inscription and manipulation – surgery, eugenics, radiography – can be reconfigured through the information economy, which has become the master economy of first-world nations.

There is, of course, no way in which the human can be separated out from these bio-techno-logical capacities. Human tissue and biomass demonstrate the same degree of responsiveness, pluripotency and waywardness to technics as do other kinds of tissue. Perhaps they demonstrate more, given the sheer intensity with which humans exploit, as well as fear, bodily variability and susceptibility to technical processes. The VHP, the Human Genome Project, tissue engineering, in-vitro fertilisation procedures, cloning – each of these plays off the openness of the human to modes of engineering and technogenesis. Each demonstrates that the point of human origin, whether phylogenic or ontogenic, the origin of species-being or of a particular organism, is susceptible to technical production, the human not as inventor but as invention. As I have demonstrated in this text, the very category 'human' owes its coherence to technologies which configure bodily morphology according to the medium-specific qualities of the archive itself – the book, the photographic archive, the computer archive and so forth. The human body as a norm of species-being can only be isolated out from the processes of the world and other organisms through a thoroughgoing technological specification, which in turn involves a complete contamination of the human with its techniques of analysis.

The VHP is an exemplary text of technogenesis. It recapitulates that most telling of human technogenic narratives, *Frankenstein,* and makes visible the extent to which the human is produced through encounters with those things that it putatively excludes – code, the corpse, its own endosoma, the computer. The VHP makes the concept of the posthuman visualisable, in that it confronts the human with those realms of production which it excludes from its self-image, showing the debt it owes to all those inhuman capacities which form its borders. The VHP lends an iconography to the idea of the human as synthetic, not a self-origin but rather the product of inestimable and incremental techno-bio-social processes.

At the same time the VHP stands at the limit of biotechnology's regenerative powers. In its reliance upon corpses it gives tacit acknowledgment to the waywardness of living process, its susceptibility not only to the ordering of techniques but to modes of dis-assemblage and disordering, those associated with the processes of death. While corpse tissue may provide medicine with a resource for processes of regeneration, and legally dead bodies may donate their residual forms of vivacity to the still-living, these ways of generating biovalue and engineering life do not provide a limitless support for the living subject. The death which medicine holds off cannot be postponed indefinitely. While biotechnology may be able to utilise the vivacity of the corporeal fragment, it cannot indefinitely maintain the corporeal integration necessary to support subjective life and consciousness. At a certain point in the body's entropic movement, subjectivity drains away, beyond a point where it can be recaptured by technical intervention. Medicine's commitment to proliferation, to vivification, to regeneration and reanimation, to an *excess* of life, derives, I think, from a desire to compensate for this limit, a multiplication of life in order to obscure the subtraction of death. The Visible Human figures have found such a central place in the biomedical imaginary precisely because they embody both this limit and its apparent transcendence, the spectacular reanimation of the corpse.

NOTES

1 THE VISIBLE HUMAN PROJECT: AN INITIAL HISTORY

1. http://www.nlm.nih.gov/research/visible/.
2. These headlines appeared, respectively, in *The Vindicator*, 28 November 1994, *The Times*, 29 November 1994 and *The Plain Dealer*, 21 November 1994.
3. *Life*, February 1997.
4. The Neuromancer series of novels by W. Gibson comprises *Neuromancer* (1984, New York: Ace Books), *Count Zero* (1987, New York: Ace Books) and *Mona Lisa Overdrive* (1995, London: HarperCollins).
5. The obvious reference here is to Lewis Carroll's *Alice Through the Looking-Glass*, whose tale of a world subject to inverted laws of time and matter prefigures the contemporary virtual imagination with wonderful precision. For a recent (1998) cinematic reworking of this theme in relation to that other pervasive screen technology, television, see *Pleasantville*.
6. This is Weber's (1998) translation of a phrase in Nietzsche's *Birth of Tragedy*.
7. Some technical aspects of these different imaging modalities are discussed later in this chapter.
8. In the process of giving numerous papers about the VHP, I have been struck again and again by the difficulty for even highly specialised academic audiences of articulating the compelling nature of these images. Any visual presentation inevitably elicits enormous interest and a kind of stunned astonishment, yet questions and comments rarely move beyond requests for more information about the Project itself. This book is thus an attempt to articulate my own sense of frustrated fascination.
9. 'About the Center for Human Simulation', http://www.uchsc.edu/sm/chs/about.html.
10. Personal communication, Dr Michael Ackerman.
11. The VHP site has hotlinks to the sites mentioned here.
12. For a discussion of this process of domestication see Chapter 2, p. 50 regarding Hartouni's (1997) account of the anxieties around 'test-tube babies' in the early 1980s, and the eventual domestication of IVF technologies.

2 POSTHUMAN SPECTACLE

1. This technical move is discussed at greater length in Chapter 4.
2. These corrections, and the problems that they in turn introduce, will be discussed at length in Chapters 3 and 4.

NOTES

3 I am using the term 'Man' deliberately, not because I accept the designation as a generic term for humanity, but rather in order to register the masculinist bias within humanism which privileges the masculine subject as the exemplar of the fully human.
4 Translator's footnote in Heidegger (1977) fn.8, p. 120.
5 This point is treated in detail in Chapter 6.
6 I am using the term macro-anatomy here to distinguish the instrumentalisation under discussion from that of micro-anatomy, and the possible instrumentations made available by genetic information for example.
7 DNA takes the form of two intertwined strands, the famous double helix, and each strand is composed of a string of chemical subunits, called nucleotides. The nucleotides have four different bases which bond into pairs and form the variable part of the structure of the helix. For a detailed account of the genetics of the Human Genome Project see Kevles and Hood (1992).
8 The term 'The Book of Man' was used by Lennart Philipson, director general of the European Molecular Biology Laboratory, to describe the Human Genome Project (cited in Kevles 1992: 30). In Chapter 3 the early anatomical idea of the body as a book is discussed at some length.
9 The computer's specific capacities for visualisation are discussed in Chapter 4.
10 One of the bioethical dilemmas presented by the HGP is that the knowledge it generates lends itself to the design of neonatal and prenatal screening for genetic disease, while treatment design is much more elusive.
11 For an account of this border control, see my study of AIDS, Waldby (1996).
12 The derivation is from ancient Greek.
13 Not surprisingly the VHP has been interpreted in these terms by some enthusiasts.
14 I am summarising a fascinating argument made by Pearson (1997) regarding the resistance of mainstream biology to evidence of the importance of symbiosis, viral cross-infection and other kinds of non-sexual, trans-species forms of genetic exchange in the process of evolution. This evidence, he writes, 'continues to play a subversive role in biology since it challenges the boundaries of the organism' (Pearson 1997: 132).
15 Here I am drawing on Hughes's (1987) idea of technological systems as propelled by the logic of 'reverse salients', that is, the way that large technical systems set up the world as an environment to be brought piece by piece under the control of the system, insofar as anything beyond the logic of its own system acts as a drag, a limit to potential, a point of friction or an inefficiency. This idea was brought to my attention in Van der Ploeg (1998: 85).
16 See for example the collection in Gray (1995b).
17 Particularly the essays in Part I: The Genesis of Cyborg.

3 THEATRES OF VIOLENCE: THE ANATOMICAL SACRIFICE AND THE ANATOMICAL TRACE

1 This in no way singles out the VHP as such. It could be argued that all technologies are also histories of technology. As Heidegger observes 'Within the complex of machinery that is necessary to physics in order to carry out the smashing of the atom lies hidden the whole of physics up to now' (Heidegger 1977: 124).
2 This has been amply demonstrated in much contemporary feminist work which examines the terms through which women were excluded from public participa-

NOTES

tion in the social contract. Chasin (1995) for example demonstrates how women as 'housewives' and domestic servants have been readily assimilated to the status of servant-machine, objects which labour for subjects. Similarly Gatens (1996) describes the assimilation of women to the status of animals and hysterical bodies as symptomatic of the implicit masculinity of the 'human', and as a strategy for excluding women from public discourse:

> Our legal and political arrangements have man as the model, the centre-piece. ... It is still 'anthropos' who is taken to be capable of representing the universal type, the universal body. Man is the model and it is his body which is taken for the human body; his reason which is taken for Reason; his morality which is formalised into a system of ethics. In our relatively recent history, the strategies for silencing those who dare to speak in another voice ... are instructive ... The first is to 'animalise' the speaker; the second is to reduce her to her 'sex'. Women who step outside their allotted place in the body politic are frequently abused with terms like harpy, virago, vixen, bitch, shrew; terms that make clear that if she attempts to speak from the political body, about the political body, her speech is not recognised as human speech.
> (Gatens 1996: 24–5)

Such strategies of exclusion are not of course limited to women. Humanism is traversed by hierarchies of the more or less human, nowhere more evident than in the social Darwinist discourse which creates hierarchies of race according to putative degrees of development and distance from animals.

3 For an account of this hierarchy in AIDS medicine, see Waldby 1996. For an account of its operation in gynaecology see Kapsalis 1997.
4 The preference for executed criminals as dissection subjects probably emerges before this date, when the practice became well established. Cazort (1996) reports reference to the use of 'alien criminals' for dissections in Florence as early as 1387.
5 http://www.uke.uni-hamburg.de/institute/imdm/idv/gallery/.
6 I make a distinction here between popular and scientific culture because, as far as I can tell, there has been no equivalent celebration of Jernigan's contribution within anatomical circles, and so little distinction, at this level, between the male and female bodies. Rather the donation of both bodies is treated as a matter-of-fact procedure, along the same lines as the donation of bodies for any other scientific enterprise.
7 Literally a 'bad subject', one beyond the pale of humanist hierarchies, but also French slang for a 'bad guy', a criminal type.
8 This transformation cannot be dated precisely. I am taking my periodisation here from the conventional History of Medicine narrative, which poses the publication of Vesalius' first anatomical text, *De Humani Corporis Fabrica*, as the pivotal moment in the emergence of modern scientific medicine.
9 The architect Vitruvius advocated the design of buildings based around the ratios of the human body. The familiar anatomical figure where the body's limbs are outstretched, forming the perimeter of a circle, is known as a Vitruvian figure.
10 Marvin describes for example an anatomy text possessed by one Dr Hough of Philadelphia which he had 'bound in the thigh skin of the patient whose body had provided the occasion for his own first appearance in the medical literature.

NOTES

Two volumes, in fact, were bound in the skin of Mary Lynn ... a twenty-eight year old Irish widow dead of consumption at ... the Almshouse facility where resident physicians learned their skills on pauper's bodies' (Marvin 1994: 136).

11 The refiguring of the Web as space of anatomical theatre was recently explicated in a public interface/installation held simultaneously at the Theatrum Anatomicum, Amsterdam and the video wall at the Guggenheim Museum Soho, New York. The installation recreated and relocated seventeenth-century public dissecting lessons held in Amsterdam's Theatrum Anatomicum as a net spectacle, displayed simultaneously at the respective sites and on the Web. The event was held on 18–21 September 1998, and was accompanied by a netlinked forum entitled 'Digi Gender Social Body: Under the Knife, Under the Spell of Anesthesia'.

12 Chapter 4 gives a more extended treatment of the spatiality of the tomographic scan.

13 As Marchessault (1996) observes, medicine has sought since the nineteenth century to render the body in mathematical terms, an attempt which, since the 1950s, has taken place primarily in the realm of cybernetic code.

14 The practice of gathering genetic material from isolated, genetically distinctive populations and patenting bioactive attributes of the cell line has caused numerous controversies in recent years. In 1995 the US National Institutes of Health patented the genes of a tribal Papua New Guinean man against the wishes of his tribe, who claimed rights in their own genetic material. This claim is disallowed under current US legislation, a ruling which denies these and other indigenous and ethnic population groups any rights in profits made from their genetic material. Similar controversies have arisen in China, northern Canada and elsewhere.

15 This theme of time, death, data and mortality is taken up at length in Chapters 5 and 6.

4 VIRTUAL SURGERY: MORPHING AND MORPHOLOGY

1 At time of writing anatomy schools using the VHP data include University of California, San Diego Medical School; SUNY Stony Brook; University of Pennsylvania Medical School; Loyola University Chicago; the Johannes Gutenburg University, Germany; the Ecole Polytechnique Federale de Lausanne, Switzerland; Washington University Medical School; Columbia University; McGill University; and SUNY Health Science Center Brooklyn. Given the speed with which the VHP data are being taken up in the wealthier academies this list will be much longer by the time this text is published.

2 This research is being carried out at the Center for Human Simulation, University of Colorado, Denver. See also the Multimedia Medical Systems Digital Humans website at http://www.mmms.com/dh/.

3 Developed at the Virtual Environment for Reconstructive Surgery project, a collaboration between the Biocomputation Center at NASA and the Department of Reconstructive Surgery at Stanford University.

4 Centre for Information Enhanced Medicine, National University of Singapore.

5 See for example the article by Satava (1995) and the edited collection Satava *et al.* (1995).

6 The history of the discovery of x-rays is by now well documented. See for example Chapter 3 of Reiser (1978), Part 1 of Kevles (1997), and Chapter 5 of Cartwright (1995) for accounts.

NOTES

7 That is with the exception of the skull. Thomas Edison conducted a famous early experiment with x-rays when in 1896, at the request of William Randolph Hearst, he tried to produce a radiograph of the brain. Edison found that the x-rays were deflected from the dense bones of the skull, and it was not until the 1920s that reliable techniques were developed for the imaging of the brain.
8 X-rays were initially restricted in their applications by the limited kinds of tissue density they registered – they traversed most internal organs in the same way that they traversed the skin and flesh. Instead of the traditional medical problem of the body's excess opacity, the x-ray produced an excessive transparency, a reversed play of the opaque and the transparent which initially rendered it difficult to use. Physicians and radiologists began to modify the body to make it more amenable to x-ray imaging, through the use of contrast agents, barium meals and tracer agents to render the digestive tract and arterial systems opaque to the rays and hence visible in x-ray spectra. With the development of techniques to eliminate 'noise' from the image, to grid and focus the x-rays and to rationalise the production equipment during the first ten years of the technology's existence, radiographs were gradually routinised in the hospital as part of its standard diagnostic procedures for a range of conditions and therapies.
9 The creation of a sterile field around the immediate operating area.
10 Rudolph Grashey *Chirurgisch-Pathologische Röntgenbilder*, 1905.
11 The cryomacrotome technique in fact has a long history of use for both animal and human anatomy, using various media – photography, cinema and more recently digital photography. Nevertheless, as I noted in the first chapter, so far only the Visible Human Project has produced an exscription of entire human bodies, as distinct from particular organs or body sections.
12 See the Multimedia Medical Systems Digital Humans website at http://www.mmms.com/dh/.
13 It is important to note that the distinction between the mutability of the digital image and the mutability of the photograph is of a relative rather than an absolute kind. While Mitchell (1992) stresses the divergence between the two media, Lister (1995) points out that in many ways the digitisation of photographic images simply enables the automation of many photomontage techniques which would previously have needed to be carried out by hand or through some mechanical or chemical process. The photograph has never, he argues, been able to lay claim to a flawless indexicality or referentiality, because the photograph has always been mutable.

5 IATROGENESIS: DIGITAL EDEN AND THE REPRODUCTION OF LIFE

1 For an analysis of the position of female anatomy and its relegation to reproductive anatomy in the VHP see Cartwright (1998).
2 'Virtual Eve joins Virtual Adam' *Washington Post* 5 December 1995; 'Adam Unterm Eishobel' *Der Spiegel* August 1994; 'A Digital Adam' *Fortune* 1 November 1993; 'Adam's Family Values' *The Economist* 5 March 1994.
3 *The Economist* 5 March 1994: 97.
4 Personal communication, Dr Victor Spitzer.
5 A case in point here is the reproduction of the OncoMouse, a genetically engineered species patented by DuPont. The patent specified that a fee must be paid after a certain number of offspring were produced, even if the offspring did not reproduce the desired characteristic. In this way DuPont secured its interests in

the OncoMouse as commodity form over the unforeseeable mutation introduced by sexual reproduction. For more on the OncoMouse as commodity and patent see Vasseleu (1996). I am particularly indebted to this article at a number of points in this chapter.
6 For an exhaustive critique of the preformationist reductionism of much genetic thought see Oyama (1985).
7 The displacement of the maternal body with technical expertise and the biomedical drive to appropriate reproductive power has been well documented and theorised by feminist intellectuals over recent years. See for example Doane (1990: 163–76).
8 Personal communication, Dr Victor Spitzer.
9 M. Saunders and C. O'Malley (eds) *The Illustrations from the Works of Andreas Vesalius of Brussels*, Cleveland, World Publishing, 1950.
10 This habit of vitality in anatomical images is so well established that the work of John Bell, an eighteenth-century anatomist whose depictions are clearly corpses, looks shocking and violent. See Roberts (1996: 71–102).
11 My thanks to Stefan Helmreich here for sending me his excellent papers and bringing this point to my attention.
12 See for example V. Kiernan, 'A Slice of Life on the Internet' *New Scientist* 3 December 1994: 5; M. Waldorp, 'The Visible Man Steps Out' *Science* 269, 1995: 1358; J. Miller, 'Anatomy Via the Internet' *Bioscience* 44, 1994: 397; Stix, 'Habeas Corpus', 1993; K. Wirthlin, 'The Visible Woman' http://www.sqi.com/winter_96/ vis_woman.html.1996.
13 These headlines were used by *The Daily Telegraph* 25 July 1995; *The Sun* 21 March 1995; and *The Columbus Dispatch* 1 December 1995, respectively.
14 Ward, B. 'Executed killer enjoys immortality on CD-Rom' *The Daily Telegraph* 25 July 1995.
15 M. Marchione 'A slice of life' *Milwaukee Journal Sentinel* 18 December 1995.
16 Due to the dessication of the frozen bodies, the residue left after the cryosectioning procedure is in fact a dust-like substance.
17 For example, see Van der Ploeg (1998) Chapters 2 and 3 for an analysis of the temporal configurations of new reproductive technologies.
18 The word 'system' comes from the Greek verb *histanai*, 'to cause to stand'.
19 For an account of complex and non-complex models in immunology see Waldby (1996) Chapter 3.
20 This term is used by Lippit (1996) in his analysis of an earlier form of medical vision, the x-ray.

6 REVENANTS: DEATH AND THE DIGITAL UNCANNY

1 The medical investment in the gaze has a huge literature. See for example Stafford (1993) and Jordanova (1989).
2 For a further usage of the idea of the biomedical imaginary see my book *AIDS and the Body Politic* (Waldby 1996).
3 For a witty and incisive study of medicine's anxious management of gynaecology images to ensure their scientific rather than pornographic interpretation see Kapsalis (1997) Chapter 4.
4 For accounts of dramatic hysterical afflictions see Kirby (1997) and McDougall (1989).
5 For a discussion of differences among immunological models see my book *AIDS and the Body Politic* (Waldby 1996) Chapter 3.

NOTES

6 There are of course medical practitioners who have productively questioned the mechanical model of the body in medicine, from a number of perspectives. Sigmund Freud and Oliver Sacks are famous examples of clinicians who take the dynamism of body and lived world seriously. Specialists working in areas of medicine more influenced by systems theory, like immunologists and geneticists, for example, are less likely to subscribe to a mechanical idea of the body and more likely to take account of the kinds of dynamic complexity modelling associated with informatic technologies.

7 Of course the micro-anatomy of death is not at all stable, as cell structures break down and succumb to bacterial processes, a development which manifests at the level of macro-anatomy pretty quickly!

8 As many commentators (Bloom 1965; Gilbert and Gubar 1979) have pointed out, *Frankenstein* is an inverted, post-lapsarian reading of the book of Genesis.

9 For an more extensive reading of the VHP through the Frankenstein narrative see Kember (1998). A recent (1997–8) exhibition on the Frankenstein story at the History of Medicine Library, a branch of the National Library of Medicine, used VHP images as contemporary manifestations of the narrative, in the face of some NLM disapproval and controversy.

10 The term cyberpunk refers to a science fiction genre which developed in the 1980s, drawing on the computer and virtual space as a narrative topography. William Gibson's *Neuromancer Trilogy* is the most famous.

11 The idea of virtual space as a site of incarceration is explored in the film *Virtuosity*, where Russell Crowe plays a criminal construct, a digital amalgam of criminal personalities used by the police force to train rookies. The construct escapes from virtual space and wreaks predictable havoc in the real world.

BIBLIOGRAPHY

Ackerman, Michael (1991) 'The Visible Human Project', *Journal of Biocommunication* 18 (2): 14.
——(1992) 'The Visible Human Project of the National Library of Medicine', *Medinfo* 92: 366–70.
Armstrong, David (1983) *Political Anatomy of the Body: Medical knowledge in Britain in the twentieth century*, Cambridge: Cambridge University Press.
Barker, Francis (1984) *The Tremulous Private Body*, London and New York: Methuen.
Bauman, Zygmunt (1992) *Mortality, Immortality and other Life Strategies*, Oxford: Blackwell.
Benedikt, Michael (ed.) (1994) *Cyberspace: First steps*, Cambridge, Mass. and London: MIT Press.
Bloom, Harold (1965) 'Afterword', *Frankenstein*, New York and Toronto: New American Library.
Boss, Pete (1986) 'Vile bodies and bad medicine', *Screen* 27 (1): 14–24.
Bowersox, J., Cordts, P. and LaPorta, A. (1998) 'Use of an intuitive telemanipulator system for remote trauma surgery: an experimental study', *Journal of the American College of Surgeons* 186 (6): 615–21.
Braidotti, Rosi (1994) 'Body-images and the pornography of representation', in K. Lennon and M. Whitford (eds) *Knowing the Difference: Feminist perspectives in epistemology*, London and New York: Routledge.
——(1997) 'Meta(l)morphoses', *Theory, Culture and Society* 14 (2): 67–80.
Bynum, Caroline Walker (1995) *The Resurrection of the Body in Western Christianity*, New York: Columbia University Press.
Canguilhem, Georges (1992) 'Machine and organism', in J. Crary and S. K.Winter (eds) *Incorporations*, New York: Zone Books.
——(1994) *A Vital Rationalist: Selected writings from Georges Canguilhem*, ed. F. Delaporte, trans. A. Goldhammer, New York: Zone Books.
Cartwright, Frederick (1967) *The Development of Modern Surgery*, London: Arthur Barker Ltd.
Cartwright, Lisa (1995) *Screening the Body: Tracing medicine's visual culture*, Minneapolis and London: University of Minnesota Press.
——(1997) 'The Visible Man: the male criminal subject as biomedical norm', in J. Terry and M. Calvert (eds) *Processed Lives: Gender and technology in everyday life*, London and New York: Routledge.

——(1998) 'A cultural anatomy of the Visible Human Project', in Paula Treichler, Constance Penley and Lisa Cartwright (eds) *The Visible Woman: Imaging technologies, gender and science*, New York: New York University Press.

Cartwright, L. and Goldfarb, B. (1992) 'Radiography, cinematography and the decline of the lens', in J. Crary and S. K. Winter (eds) *Incorporations*, New York: Zone Books.

Casper, Monica (1995) 'Fetal cyborgs and technomoms on the reproductive frontier: which way to the carnival?', in C. Gray (ed.) *The Cyborg Handbook*, New York and London: Routledge.

Cazort, Mimi (1996) 'The theatre of the body', in Mimi Cazort, Monique Kornell and K.B. Roberts (eds) *The Ingenious Machine of Nature: Four centuries of art and anatomy*, Ottawa: National Gallery of Canada.

Chasin, Alexandra (1995) 'Class and its close relations: identities among women, servants and machines', in J. Halberstram and I. Livingston (eds) *Posthuman Bodies*, Bloomington and Indianapolis: Indiana University Press.

Cixous, Hélène (1976) 'Fiction and its phantoms: a reading of Freud's Das Unheimliche', *New Literary History* 7: 525–48.

Crawford, T. Hugh (1996) 'Imaging the human body: quasi objects, quasi texts, and the theatre of proof', *PMLA* 3 (1): 66–79.

Csordas, Thomas (1998) 'Computerised cadavers: shades of being and representation in virtual reality', unpublished manuscript.

Cubitt, Sean (1996) 'Supernatural futures: theses on digital aesthetics', in G. Robertson, M. Mash, L. Tickner, J. Bird, B. Curtis and T. Putnam (eds) *FutureNatural: Nature, science, culture*, London and New York: Routledge.

Dagognet, François (1988) *La Maîtresse du vivant*, Paris: Hachette.

Daston, Lorraine and Galison, Peter (1992) 'The image of objectivity', *Representations* 40: 81–128.

De Landa, Manuel (1997) *A Thousand Years of Non-Linear History*, New York: Zone Books.

Deleuze, Gilles (1990) *The Logic of Sense*, trans. Mark Lester, New York: Columbia University Press.

DeLillo, Don (1985) *White Noise*, New York: Penguin.

Derrida, Jacques (1989) 'Psyche: inventions of the other', in Lindsey Waters and Wlad Godzich (eds) *Reading de Man Reading*, Minneapolis: University of Minnesota Press.

——(1994) *Specters of Marx. The state of debt, the work of mourning and the New International*, trans. Peggy Kamuf, New York and London: Routledge.

Doane, Mary-Anne (1990) 'Technophilia: technology, representation and the feminine', in M. Jacobus, E.F. Keller and S. Shuttleworth (eds) *Body/Politics. Women and the discourses of science*, New York and London: Routledge.

Dowling, C. (1997) 'How one death-row prisoner found afterlife on the internet', *Life*, February: 40–7.

Edwards, Paul (1990) 'The army and the micro-world: computers and the politics of gender identity', *Signs* 16 (1):102–27.

Edwards, P., Hill, D., Hawks, D. (1996) 'The Visible Human dataset as an atlas and a source of test data for model-based surgery guidance', in *The Visible Human Project Conference Proceedings*, 7–8 October, National Library of Medicine, Bethesda, Md.

BIBLIOGRAPHY

Eisenstein, Elizabeth (1979) *The Printing Press as an Agent of Change: Communications and cultural transformations in Early-Modern Europe*, Volumes I and II, Cambridge: Cambridge University Press.

Ellison, David (1995) 'Anatomy of a murderer', *21.C* 3: 20–5.

Forbes, Thomas (1981) 'To be dissected and anatomised', *Journal of the History of Medicine and Allied Sciences* 36: 490–2.

Foucault, Michel (1972) *The Archaeology of Knowledge*, trans. A.M.S. Smith, London: Tavistock.

——(1975) *The Birth of the Clinic: An archeology of medical perception*, New York: Vintage Books.

——(1979) *Discipline and Punish*, Harmondsworth, Middx: Peregrine Books.

——(1980) *The History of Sexuality. Volume I: An introduction*, New York: Vintage Books.

Franklin, Sarah and Ragoné, Helena (1998) (eds) *Reproducing Reproduction: Kinship, power and technological innovation*, Philadelphia: University of Pennsylvania Press.

Freud, Sigmund (1919) 'The uncanny', *The Complete Works of Sigmund Freud*, Standard Edition, vol. 17: 219–52.

Gatens, Moira (1996) *Imaginary Bodies: Ethics, power and corporeality*, London and New York: Routledge.

Gilbert, Sandra and Gubar, Susan (1979) *The Madwoman in the Attic*, New Haven and London: Yale University Press.

Gray, C. (1995a) 'Science fiction becomes military fact' in Gray, C. (ed.) *The Cyborg Handbook*, New York and London: Routledge.

——(ed.) (1995b) *The Cyborg Handbook*, New York and London: Routledge.

Grosz, Elizabeth (1994) *Volatile Bodies: Toward a corporeal feminism*, Sydney: Allen and Unwin.

——(1995) *Space, Time and Perversion: The politics of bodies,* Sydney: Allen and Unwin.

Halberstram, J. and Livingston, I. (1995) (eds) *Posthuman Bodies*, Bloomington and Indianapolis: Indiana University Press.

Haraway, Donna (1991) *Simians, Cyborgs and Women. The reinvention of nature*, New York: Routledge.

Harcourt, Glenn (1987) 'Andreas Vesalius and the anatomy of antique sculpture', *Representations* 17: 28–61.

Hartfield, Scott (1995) 'Visible man to prove more than just an excellent study of anatomy', *Advance* 24 April.

Hartouni, Valerie (1997) *Cultural Conceptions: On reproductive technologies and the remaking of life*, Minneapolis and London: University of Minnesota Press.

Hayles, N. Katherine (1993) 'Virtual bodies and flickering signifiers', *October* 66, Fall: 69–91.

——(1995) 'The life cycle of cyborgs: writing the posthuman', in C. Gray (ed.) *The Cyborg Handbook*, New York and London: Routledge.

——(1996) 'Narratives of artificial life', in G. Robertson, M. Mash, L. Tickner, J. Bird, B. Curtis and T. Putnam (eds) *FutureNatural: Nature, science, culture*, London and New York: Routledge.

Heelan, Patrick (1983) *Space Perception and the Philosophy of Science,* Berkeley: University of California Press.

Heidegger, Martin (1977) *The Question Concerning Technology and Other Essays*, translated and introduced by William Lovitt, New York: Harper & Row.

Heim, M. (1994) 'The erotic ontology of cyberspace', in M. Bendickt (ed.) *Cyberspace: First steps*, Cambridge, Mass and London: MIT Press.

Helmreich, Stefan (1994) 'Anthropology inside and outside the looking-glass worlds of Artificial Life'. Online. Available HTTP: htpp://www.gslis.utexas.edu (8 April 1994).

——(1998) 'Replicating reproduction in artificial life: or, the essence of life in the age of virtual electronic reproduction', in S. Franklin and H. Rogoné (eds) *Reproducing Reproduction*, Philadelphia: University of Pennsylvania Press.

Hirschauer, S. (1991) 'The manufacture of bodies in surgery', *Social Studies of Science* 21: 279–319.

Hogle, Linda (1995) 'Tales from the cryptic: technology meets organism in the living cadaver', in C. Gray (ed.) *The Cyborg Handbook*, New York and London: Routledge.

Hohne, K.H., Pflesser, B., Pommert, A., Riemer, M., Schiemann, T., Schubert, R. and Tiede, U. (1996) 'A virtual body model for surgical education and rehearsal', *Computer* January: 25–31.

Hong, L., Kaufman, A., Wei, Y.-C., Viswambharan, A., Wax, M. and Liang, Z. (1995) '3D virtual colonoscopy', in *Proceeds of the IEEE Symposium on Frontiers of Biomedical Visualisation*, Los Alamitos, IEEE CS Press: 26–32.

Hong, L., Kaufman, A., Liang, Z., Viswambharan, A. and Wax, M. (1996) 'Visible Human virtual colonoscopy', in *The Visible Human Project Conference Proceedings*, 7 and 8 October 1996, National Library of Medicine, Bethesda, Md.

Hood, L. (1992) 'Biology and medicine in the twenty-first century', in D. Kevles and L. Hood (eds) *The Code of Codes: Scientific and social issues in the Human Genome Project*, Cambridge, Mass: Harvard University Press.

Howell, Joel (1995) *Technology in the Hospital: Transforming patient care in the early twentieth century*, Baltimore and London: Johns Hopkins University Press.

Hughes, T. (1987) 'The evolution of large technological systems', in W. Bijker, T. Hughes and T. Pinch (eds) *The Social Construction of Technological Systems. New directions in the sociology and history of technology*, Cambridge, Mass: MIT Press.

Hutchings, Peter (1997) 'Spectacularising crime: ghostwriting the law', *The 15th Annual Law and Society Conference*, Keynote Address, La Trobe University, Melbourne, Australia, 3–5 December.

Idhe, Don (1983) 'The historical-ontological priority of technology over science', in P.T. Durbin and F. Rapp (eds) *Philosophy and Technology*, Reidel: Dordrecht.

Innis, Robert (1984) 'Technics and the bias of perception', *Philosophy and Social Criticism* 10 (1): 67–89.

Jonson, A, (1999) 'Still platonic after all these years: artificial life and form/matter', *Australian Feminist Studies* 14, 29: 47–61.

Jordanova, Ludmilla (1989) *Sexual Visions: Images of gender in science and medicine between the eighteenth and twentieth centuries*, University of Wisconsin Press.

Kapsalis, Terri (1997) *Public Privates: Performing gynecology from both ends of the speculum*, Durham and London: Duke University Press.

Keane, Helen (1997) 'What's wrong with addiction?', Doctoral Thesis, Australian National University, Canberra.

Keller, E.F. (1992) 'Nature, nurture and the Human Genome Project', in D. Kevles and L. Hood (eds) *The Code of Codes: Scientific and social issues in the Human Genome Project*, Cambridge, Mass: Harvard University Press.

BIBLIOGRAPHY

——(1995) *Refiguring Life: Metaphors of twentieth-century biology*, New York: Columbia University Press.

——(1996) 'The biological gaze', in G. Robertson, M. Mash, L. Tickner, J. Bird, B. Curtis and T. Putnam (eds) *FutureNatural: Nature, science, culture*, New York and London: Routledge.

Kember, Sarah (1991) 'Medical diagnostic imaging: the geometry of chaos', *New Formations* 15: 55–66.

——(1998) *Virtual Anxiety: Photography, new technologies and subjectivity*, Manchester: Manchester University Press.

Kerr, J., Sellberg, M., Ratiu, P., Knapp, D. and Caon, C. (1996) 'Photorealistic volume rendered anatomical atlases and interactive virtual dissections of the dissectable human', in *The Visible Human Project Conference Proceedings*, 7–8 October 1996, National Institutes of Health, Bethesda, Md USA.

Kevles, Bettyann (1997) *Naked to the Bone: Medical imaging in the twentieth century*, New Brunswick: Rutgers University Press.

Kevles, D. (1992) 'Out of eugenics: the historical politics of the human genome', in D. Kevles and L. Hood (eds) *The Code of Codes: Scientific and social issues in the Human Genome Project*, Cambridge, Mass: Harvard University Press.

Kevles, D. and Hood, L. (eds) (1992) *The Code of Codes: Scientific and social issues in the Human Genome Project*, Cambridge, Mass: Harvard University Press.

Kiernan, Vincent (1994) 'A slice of life on the Internet', *New Scientist* 3 December: 5.

Kirby, Vicki (1997) *Telling Flesh: The substance of the corporeal*, New York and London: Routledge.

Kristeva, Julia (1982) *The Powers of Horror. An essay on abjection*, trans. Leon Roudiez, New York: Columbia University Press.

Langton, C. (1996) 'Artificial Life', in M. Boden (ed.) *The Philosophy of Artificial Life*, Oxford: Oxford University Press.

Latour, Bruno (1990) 'Drawing things together' in M. Lynch and S. Woolgar (eds) *Representation in Scientific Practice*, Cambridge, Mass: MIT Press.

——(1993) *We Have Never Been Modern*, trans. Catherine Porter, Cambridge, Mass: Harvard University Press.

Leder, Drew (1990) *The Absent Body,* Chicago: University of Chicago Press.

——(1992) 'A tale of two bodies: the Cartesian corpse and the lived body', in Drew Leder (ed.) *The Body in Medical Thought and Practice*, Dordrecht, Boston and London: Klewer Academic Publishers.

Le Doeuff, Michelle (1989) *The Philosophical Imaginary*, trans. C. Gordon, Stanford: Stanford University Press.

Lippit, Akira Mizuta (1996) 'Phenomenologies of the surface: radiation-body-image', *Qui Parle* 9 (2): 31–50.

——(1998) 'Radiant images, radioactive bodies', Catalogue for the 44th International Short Film Festival, Oberhausen: 126–30.

Lister, Martin (1995) 'Introductory essay', in M. Lister (ed.) *The Photographic Image in Digital Culture*, New York and London: Routledge.

Lorensen, W., Jolesz, F. and Kikinis, R. (1995) 'The exploration of cross-sectional data with a virtual endoscope', in R. Satava, K. Morgan, H. Sieburg, R. Mattheus and J. Christensen (eds) (1995) *Interactive Technology and the New Paradigm for Healthcare*, Amsterdam: IOS Press.

Lynch, M (1988) 'Sacrifice and the transformation of the animal body into a scientific object: laboratory culture and ritual practice in the neurosciences', *Social Studies of Science* 18: 265–89.

Lynch, M. and Woolgar, S. (1990) 'Introduction: sociological orientations to representational practice in science', in M. Lynch and S. Woolgar (eds) *Representation in Scientific Practice*, Cambridge, Mass: MIT Press.

McDougall, Joyce (1989) *Theatres of the Body: A psychoanalytic approach to psychosomatic illness*, London: Free Association Books.

McNamee, T. (1994) 'Computerised cadaver proves a visible success', *Chicago Sun-Times* 29 November.

Marchessault, Janine (1996) 'The secret of life: informatics and the popular discourse of the life code', *New Formations* 19: 120–9.

Marvin, Carolyn (1994) 'The body of the text: literacy's corporeal constant', *The Quarterly Journal of Speech* 80: 129–49.

Masys, Daniel (1990) 'Picture this ... Developing standards for electronic images at the National Library of Medicine', *SCAMC Inc*: 20–1.

Meglan, D. (1996) 'Making surgical simulation real', *Computer Graphics* 30 (4): 37–9.

Miller, Julie (1994) 'Anatomy via the Internet', *Bioscience* 44: 397.

Mitchell, William (1992) *The Reconfigured Eye: Visual truth in the post-photographic era*, Cambridge, Mass: MIT Press.

Moore, L. and Clarke, A. (2000) 'The traffic in cyberanatomies: sex/gender/sexualities in local and global formations', *Body and Society*, forthcoming.

Muller, J. and Koenig, B. (1988) 'On the boundaries of life and death: the definition of dying by medical residents', in M. Lock and D. Gordon (eds) *Biomedicine Examined*, Dordrecht, Boston and London: Klewer Academic Publishers.

Myers, Greg (1990) 'The double helix as icon', *Science as Culture* 9: 49–72.

Oyama, Susan (1985) *The Ontogeny of Information: Developmental systems and evolution*, Cambridge: Cambridge University Press.

Pearson, Keith Ansell (1997) *Viroid Life: Perspectives on Nietzsche and the transhuman condition*, London and New York: Routledge.

Petchesky, Rosalind (1987) 'Fetal images: the power of visual culture in the politics of reproduction', *Feminist Studies* 13 (2): 263–92.

Playter, R. and Raibert, M. (1997) 'A virtual surgery simulator using advanced haptic feedback', *Minimally Invasive Therapy and Allied Technology* 6: 117–21.

Pommert, A., Bomans, M., Riemer, M., Tiede, U. and Höhne, K. (1993) '3-D imaging in medicine', in A. Kent and J. Williams (eds) *Encyclopedia of Computer Science and Technology* vol. 28, supplement 13, New York: Marcel Dekker Inc.

Rabinow, Paul (1996) *Essays on the Anthropology of Reason*, Princeton, NJ: Princeton University Press.

Race, Kane (1999) 'Incorporating clinical authority: a new test for people with HIV', in N. Watson (ed.) *Reframing Bodies*, Melbourne, Australia: Macmillan.

Reiser, Stanley (1978) *Medicine and the Reign of Technology*, Cambridge: Cambridge University Press.

Richardson, Ruth (1988) *Death, Dissection and the Destitute*, London: Routledge and Kegan Paul.

Robb, Richard (1996) 'Virtual endoscopy: development and evaluation using the Visible Human datasets', in *The Visible Human Project Conference Proceedings*, 7–8 October, National Library of Medicine, Bethesda, Md.

Roberts K. (1996) 'The contexts of anatomical illustration', in M. Cazort, M. Kornell and K.B. Roberts (eds) *The Ingenious Machine of Nature: Four centuries of art and anatomy*, Ottawa: National Gallery of Canada.

Robins, Kevin (1996) *Into the Image: Culture and politics in the field of vision*, London and New York: Routledge.

Ross, M. (1996) '3-D imaging in virtual environment: a scientific, clinical and teaching tool', in *The Visible Human Project Conference Proceedings*, 7–8 October 1996, National Library of Medicine, Bethesda, Md.

Satava, R. (1995) 'Computers in surgery: telesurgery, virtual reality and the new world order of medicine', *Contemporary Surgery* 47 (4): 204–8.

Satava, R., Morgan, K., Sieburg, H., Mattheus, R. and Christensen, J. (eds) (1995) *Interactive Technology and the New Paradigm for Healthcare*, Amsterdam: IOS Press.

Saunders, M. and O'Malley, C. (eds) (1950) *The Illustrations from the Works of Andreas Vesalius of Brussels*, Cleveland: World Publishing Company.

Sawday, Jonathan (1995) *The Body Emblazoned: Dissection and the human body in Renaissance culture*, London and New York: Routledge.

Schiebinger, Londa (1987) 'Skeletons in the closet: the first illustrations of the female skeleton in eighteenth-century anatomy', in C. Gallagher and T. Laqueur (eds) *The Making of the Modern Body: Sexuality and society in the nineteenth century*, Berkeley: University of California Press.

Schrödinger, Erwin (1944) *What is Life? The physical aspect of the living cell*, London: Cambridge University Press.

Serres, Michel (1982) *Hermes: Literature, science, philosophy*, ed. J.V. Harari and D.F. Bell, Baltimore: John Hopkins University Press.

Shapin, S. and Schaffer, S. (1985) *Leviathan and the Air-Pump: Hobbes, Boyle, and the experimental life: including a translation of Thomas Hobbes, 'Dialogus physicus de natura aeris'*, Princeton, NJ: Princeton University Press.

Shelley, Mary (1818) *Frankenstein, or the Modern Prometheus*, London: Everyman Library, 1977.

Shildrick, M. (1996) 'Posthumanism and the monstrous body', *Body and Society* 2 (1): 1–15.

Siegel, C. (1995) 'Creating 3D models from medical images using AVS: a user's perspective', *Computer Graphics* 29 (2): 59–60.

Sobchack, Vivian (1997) 'Meta-Morphing', *Art and Text* 58: 43–5.

Sochurek, Howard (1987) 'Medicine's new vision', *National Geographic* 171, 1 (January): 2–41.

Spitzer, V., Ackerman, M., Scherzinger, A. and Whitlock, D. (1996) 'The Visible Human Male: a technical report', *Journal of the American Informatics Association* 3 (2): 118–30.

Stafford, Barbara (1993) *Body Criticism: Imagining the unseen in Enlightenment art and medicine*, Cambridge, Mass and London: MIT Press.

Starr, Paul (1982) *The Social Transformation of American Medicine*, New York: Basic Books.

Stengers, Isabelle and Prigogine, Ilya (1997) 'The re-enchantment of the world', in

Isabelle Stengers (ed.) *Power and Invention: Situating science*, Minneapolis: University of Minnesota Press.

Stix, Garry (1993) 'Habeas Corpus: seeking subjects to be a digital Adam and Eve', *Scientific American* 268 (1): 122–3.

Stone, Allucquere Rosanne (1992) 'Virtual systems', in J. Crary and S.K Winter (eds) *Incorporations*, New York: Zone Books.

Terranova, Tiziana (1996) 'Posthuman unbounded: artificial evolution and high-tech subcultures', in G. Robertson, M. Mash, L. Tickner, J. Bird, B. Curtis and T. Putnam (eds) *FutureNatural: Nature, Science, Culture*, London and New York: Routledge.

Tiede, U., Schiemann, T. and Höhne, K. (1996) 'Visualising the Visible Human', *IEEE Computer Graphics and Applications* 16(1): 7–9.

Tierney, Thomas (1997) 'Death, medicine and the right to die: an engagement with Heidegger, Bauman and Baudrillard', *Body and Society* 3 (4): 51–77.

Tomas, David (1994) 'Old Rituals for New Space: *Rites de passage* and William Gibson's cultural model of cyberspace', in M. Bendikt (ed.) *Cyberspace: First steps*, Cambridge, Mass and London: MIT Press.

——(1995) 'Feedback and cybernetics: reimaging the body in the age of cybernetics', in M. Featherstone and R. Burrows (eds) *Cyberspace, Cyberbodies, Cyberpunk: Cultures of technological embodiment*, London: Sage.

Treichler, P., Penley, C. and Cartwright, L. (eds) (1998) *The Visible Woman: Imaging technologies, gender and science*, New York: New York University Press.

Van der Ploeg, Irma (1998) 'Prosthetic bodies: female embodiment in reproductive technologies', Doctoral thesis, University of Maastricht, Netherlands.

Vasseleu, Cathryn (1996) 'Patent pending: laws of invention, animal life forms and bodies as ideas', in P. Cheah, D. Fraser and J. Grbich (eds) *Thinking Through the Body of the Law*, Sydney: Allen and Unwin.

Waldby, Catherine (1996) *AIDS and the Body Politic: Biomedicine and sexual difference*, London and New York: Routledge.

——(1997a) 'The body in the digital archive: the Visible Human Project and the computerisation of medicine', *Health: An Interdisciplinary Journal for the Social Study of Health, Illness and Medicine* 1 (2): 227–43.

——(1997b) 'Revenants: the Visible Human Project and the digital uncanny', *Body and Society* 3 (1): 1–16.

Waldorp, M. Mitchell (1995a) 'On-line archives let biologists interrogate the genome', *Science* 269 (8 September): 1356–8.

——(1995b) 'The Visible Man steps out', *Science* 269: 1358.

Wark, McKenzie (1993) 'Lost in space: in the digital image labyrinth', *Continuum* 7 (1): 140–60.

Warwick, R. and Williams, P.L. (eds) (1973) *Gray's Anatomy*, 35th edition, Edinburgh: Longman.

Weber, Samuel (1996) *Mass Mediauras: Form, technics, media*, ed. A. Cholodenko, Sydney: Power Publications.

——(1998) 'Theatrocracy, democracy and digitisation', paper presented on 20 August to the School of Theatre, Film and Dance, University of New South Wales, Sydney, Australia.

Weiner, Norbert (1948) *Cybernetics: or Control and Communication in the Animal and the Machine*, New York: Wiley.

——(1968) *The Human Use of Human Beings. Cybernetics and society*, London: Sphere Books. First published 1950.

Wells, H.G. (1977) *The Invisible Man*, simplified and abridged by T.S. Gregory, London: Longman. First published 1897.

Wilkie, Tom (1996) 'Genes 'R' Us', in G. Robertson, M. Mash, L. Tickner, J. Bird, B. Curtis and T. Putnam (eds) *FutureNatural: Nature, Science, Culture*, London and New York: Routledge.

Wirthlin, Katherine (1996) 'The Visible Woman', http://www.sqi.com/winter_96/vis_woman.html (August 1996).

Wolfe, Cary (1995) 'In search of post-humanist theory: the second-order cybernetics of Maturana and Valera', *Cultural Critique* 30: 33–70.

Young, Katharine (1997) *Presence in the Flesh: The body in medicine*, Cambridge, Mass: Harvard University Press.

Zuiderveld, K. (1996) 'VR in radiology: first experiences at University Hospital Utrecht', *Computer Graphics* 30 (4): 47–8.

INDEX

abdominal surgery 92
accuracy 105–6, 107
Ackerman, M. 3, 12, 72–3
Adam and Eve 111–12; *see also* Genesis iconography
age 17–18
algorithms 123, 126
anatomical atlases 27, 62–70, 140
anatomical norm 106–8
anatomical theatre 59–60, 71, 158, 166
anatomy 20–1, 34, 51–80, 82, 157; biopolitical and biosemiotic economies 78–80; body as laminar terrain 63–6; body as readable terrain 66–70; human archives 6–7, 34, 37–42, 59–70; movement depicted in anatomical figures 117–18, 168; realism 26–7; sacrificial economies 52–9; space and virtual space 158; tomographic space and digital inscription 75–8; virtual map 70–5
Anatomy Acts (1830s) 53–4
animation 16, 17; point of view, morphability and 72–4; of vitality 116–20
anthropo-dermic bibliopegy 67, 165–6
anti-natural 152
antiretroviral therapies 113
archives, human 6–7, 34, 37–42, 59–70
Armstrong, D. 67–8
Artificial Life (A-Life) 22, 112, 123–7
atlases, anatomical 27, 62–70, 140
automata 118–19

Barker, F. 53
Bauman, Z. 142
Bell, J. 168
Benedikt, M. 4

biofeedback loops 122–3
bio-graphs 60–1
bioinformatics 45–6
biomedical imagination 136–7; death and 139–46
biomedicine *see* medicine
biopolitical economy 78–9
bio-power 33
biosemiotic economy 79–80
biotechnology 6–7, 19, 33–4, 161–2
biovalue 19, 33–4, 51–2
bit strings 123, 126
Bloom, H. 169
'Book of Man' 37, 164
books: bodies and 67; bound in human skin 67, 165–6
Boss, P. 148
Bowersox, J. 89
Braidotti, R. 115
brain death 143, 146
bricoleur 32
Bynum, C.W. 131

cadavers: donor 143, 146–7; selection for VHP 12–13; transformation into anatomical text 13–15
Canguilhem, G. 118, 121
capitalisation 61
Carroll, L. 163
Cartwright, L. 159, 166, 167; cinema 119–20; Jernigan 54, 56; surgery 84, 90; tomographic imaging 94, 95–6; Visible Woman 17–18
Casper, M. 143
catheterisation 85
causality 28
Cazort, M. 52, 112
Center for Human Simulation 12–13

INDEX

Chasin, A. 165
Christianity 131–2; *see also* Genesis iconography
Chronicle of Higher Education 54
chronograph 119
cinema 119–20
Cixous, H. 151, 155
Clarke, A. 55
cloning 50
codes 126–7; genetic code 44, 114, 124–5
colonoscopy, virtual 102–3
Coma 148
communications engineering 122
computed tomography (CT) scans 9, 10, 14, 25, 82–3, 107; image produced 74, 77, 95; and tissue 100, 101
computer imaging *see* digital images
computer programming 122
consciousness 153
construction 64–6
constructs 153
copies 133–5
corpse 22–3, 146–7; biomedical imagination and 142–6; 'called to account' by the VHP 34–5; *see also* cadavers; death
craniofacial surgery 85–6
criminals 52–4
criticism, posthuman 48–50
cryomacrotome technique 14, 96, 167
cryosectioning 14, 76–7, 129
Csordas, T. 82
CT scans *see* computed tomography
Cubitt, S. 22, 121, 151, 152
cybernatural life 120–7, 152
cybernetic turn 19, 45–6
cyberpunk fiction 153
cybersciences 121–2
cyberspace 4–5
cyborg 46–8

Da Vinci, Leonardo 2, 3, 4, 53
Dagognet, F. 32, 39
Daston, L. 57, 94–5, 98–9
data homologues 106–8
De Humani Corporis Fabrica (Vesalius) 64–6, 66, 69–70, 117
De Landa, M. 32
death 22–3, 36, 41, 136–56, 162; and the biomedical imagination 139–46; deferral of 141–2, 155–6; digital uncanny 149–56; indebtedness to time and 79–80; medical horror 146–9; medicalisation of 141–2; resurrection 128–31; VHP's impact on distinction between life and 160–1
Deleuze, G. 133–4
DeLillo, D. 6
depth 94–6; and surface 159–60
Derrida, J. 132–3, 151, 154
Descartes, R. 153
desire, IatroGenic 21–2, 113–16, 127, 136; *see also* Genesis iconography
differential specificity 59
digital images 9–12, 25; transformation of body into 13–16
digital photography 100–1
digital uncanny 149–56
dissection 55, 62, 142; public 52–4, 59–60
dissection rehearsal 74–5, 82
dissemination, public 138–9
Doane, M.-A. 168
donor cadavers 143, 146–7
double, the 150
Dowling, C. 54

economies of productivity 78–80
Edison, T. 167
Edwards, P. 120
Eisenstein, E. 60
Ellison, D. 15
emplacement 31
endoscopes 90
endoscopy 83, 103; virtual 73, 83–4, 102–6, 107
en-framing 41–2; technological frame 27–33; virtual 24–7
entropy 124
exosoma 40
exposition 69
exscription 89–98; radiographic 91–3; televising interior morphologies 89–91; tomographic 94–8

feedback: biofeedback loops 122–3; haptic feedback systems 84
Fifth Element, The 2
flaying 64
flythroughs 16, 17, 72–3; virtual surgery 101–6
foetal imaging 119
Forbes, T. 52

INDEX

Foucault, M. 52, 54, 141, 147; anatomy 24, 64, 66, 108; bio-power 33; life 118; sciences of 'Man' 40–1
framing *see* en-framing
Frankenstein (Shelley) 148–9, 169
Freud, S. 150–1, 169

Galenic medicine 59
Galison, P. 57, 94–5, 98–9
Gatens, M. 52, 165
GenBank 38
gene therapy 26
Genesis iconography 21–2, 110–35, 136, 139, 148–9; animation of vitality 116–20; cybernatural life 120–7; digital Eden 111–16; reproduction 131–5; resurrection of the body 128–31
genetic code 44, 114, 124–5
genetic research 122–3, 127
ghosts 150–1, 153–5
Gibson, W. 163, 169
Gilbert, S. 169
Goldfarb, B. 94, 95–6
Grashey, R. 167
Gray, C. 47, 123, 164
Grosz, E. 144, 159
Gubar, S. 169

haptic feedback systems 84
Haraway, D. 31, 43, 45–6, 121
Harcourt, G. 66, 117–18
Hartfield, S. 4
Hartouni, V. 50, 143, 163
Hayles, N.K. 47–8, 126, 153
Heelan, P. 31
Heidegger, M. 19, 42, 164; question of technology 27–33
Heim, M. 153, 155
Helmreich, S. 112, 123, 126, 168
Hirschauer, S. 34, 69, 93
Hogle, L. 143
Hohne, K.H. 74–5
homologues, patient-data 106–8
Hong, L. 72, 102–3
Hood, L. 7, 114, 164
horror, medical 146–9
Howell, J. 79, 89
Hughes, T. 164
human archives 6–7, 34, 37–42, 59–70
Human Genome Project 7, 19, 26, 44, 112, 114; human archive 37–42

Hutchings, P. 157
hybridisation 43–4, 45

iatrogenesis 113
IatroGenic desire 21–2, 113–16, 127, 136; *see also* Genesis iconography
Idhe, D. 27
in-vitro fertilisation (IVF) 50
indebtedness 54; economies of VHP's productivity 78–80
individuality 57
information 7, 160; bioinformatics 45–6; cybernatural life 121–7
Innis, R. 40
inscription, digital 75–8
instrumentation 28, 39–40
integers, array of 100–1
Intel Computers 3
interiority 5–6, 159–60
interpretation 137–8
intraoperative visualisation 106–8
invention 132–3; human as 50

Jernigan, Joseph Paul 1, 36–7, 52, 54, 56, 157; selection as Visible Male 13–14
Jonson, A. 124, 127
Jordanova, L. 66, 168

Kapsalis, T. 136, 165, 168
Keane, H. 141
Keller, E.F. 37, 55, 91, 121–2, 127, 133
Kember, S. 115–16, 169
Kerr, J. 27
Kevles, B. 138, 166
Kevles, D. 7, 37
Kiernan, V. 168
Kirby, V. 48, 168
knee 85
knowledge 60; scientific 27–8, 29–31
Koenig, B. 141–2
Kristeva, J. 147

laminar terrain 63–6
Langton, C. 124–5
laryngoscopes 90–1
latency 36–7
Latour, B. 27, 45, 63, 98, 109; preserving visual knowledge 61–2; scientific modernity 43–4
Le Doeuff, M. 137
Leder, D. 143–4, 145–6, 147, 153

181

INDEX

libidinal body 144–5
licences 16–17
life 21–2, 110–35, 162; animation of vitality 116–20; Artificial Life (A-Life) 22, 112, 123–7; biomedical imagination and death 144–6; cybernatural 120–7, 152; engineered vitality 155–6; reproduction 131–5; resurrection 128–31; uncanny vitality 149–56; VHP and distinction between life and death 160–1
Life magazine 2–3, 54
Lippit, A.M. 89, 91–2, 120, 159, 168
Lister, M. 167
Lorensen, W. 105
Lynch, M. 55, 56, 70

McDougall, J. 168
McNamee, T. 111
Macy Conferences 47–8
magnetic resonance imaging (MRI) 9, 11, 14, 82–3, 107; images produced 74; tissues 100, 101
mapping 63–78; body as laminar terrain 63–6; body as readable terrain 66–8; tomographic space 75–8; virtual map 70–5
Marchessault, J. 119, 166
Marchione, M. 168
marginalisation 53–4, 79
Marvin, C. 67, 165–6
Mayo brothers 92
mechanisms 130–1, 144–6
medical horror 146–9
medical imaging 5–6, 9–12, 25, 74; slippage between different modalities 86–7
medical pedagogy 25
medicalisation of death 141–2
medicine 7–8, 18, 38–9, 162; and death 22–3, 139–46, 150–1; Galenic 59; posthuman 42–50; privatised 79; representation 115; *see also* biomedical imagination
'Medicine and Virtual Reality' conference 1996 82
Meglan, D. 103
memento mori 140
microworlds 120
Miller, J. 168
minimally invasive surgery 88
Mitchell, W. 101, 102, 167

modernity, scientific 43–4
molecular biology 25–6, 122–3
Moore, L. 55
Moravec, H. 153
morbid and mortal matter 39
morphability 72–4
morphogenesis 127
morphology: and the anatomical norm 166–8; surrogacy and 98–101
movement 119–20; *see also* animation
MRI *see* magnetic resonance imaging
Muller, J. 141–2
Murder Act (1752) 52
Myers, G. 112

National Library of Medicine (NLM) 1, 11–12, 16–17, 18, 79, 138
National University of Singapore Centre for Information-enhanced Medicine 17
natural, the 121
nature: technology and 29–32, 161
negative presence 36–7
New York University School of Medicine 106
norm, anatomical 106–8

O'Malley, C. 168
OncoMouse 25, 114, 167–8
opacity 24
open surgery 88, 92–3
operative images 27, 34, 108–9
ophthalmoscopes 90
Oyama, S. 122, 125–6, 127, 168

palindromic temporality 129–31
palpation 89
particularity 57
pathology 24, 66
patient-data homologues 106–8
Pearson, K.A. 43, 45, 164
pelvic girdle model 87
perspective 63
Petchesky, R. 119
photography 100–1
photorealism 74–5
physics 29–30; quantum mechanics 31
placement 41–2
planar space 95
Playter, R. 84, 86
poiēsis 28–9, 32
points of view 72–4

INDEX

Pommert, A. 101
popularisation 138–9
pornography 114–15
posthuman 19–20, 161–2; posthuman medicine 42–50
post-natural, the 121, 152–4; *see also* cybernatural life
precision 105–6, 107
Prigogine, I. 32
printed books 60–2
privatised medicine 79
productivity 28–9; economies of VHP's productivity 78–80
programmers 126–7
prosthetic transformations 113
proximation 30–1
public dissection 52–4, 59–60
public dissemination 138–9
purification 44, 45

quantum mechanics 31

Rabinow, P. 32, 34, 37, 44, 131
Race, K. 113
radiography 89, 91–3, 94–5, 167
Raibert, M. 84, 86
Ray, T. 125
readable terrain 66–70
redemption 54, 57, 157
rehearsal 74–5
Reiser, S. 90–1, 166
representation, medical 115
reproduction 115–16; Genesis iconography 131–5; reproductive status of Visible Woman 17–18
restacking 15
resurrection 128–31, 131–2
revelation 29
revenants 150–1, 153–5
reversibility 129–31
Richardson, R. 54
Robb, R. 73, 83–4, 87, 104–5, 107
Roberts, K. 168
Robins, K. 4, 100
romanticism 32
Ross, M. 85–6

Sacks, O. 169
sacrifice 36, 70; sacrificial economies 52–9
safety standards 107
Satava, R. 166

Saunders, M. 168
Sawday, J. 57, 62, 63–4, 64–6, 158; anatomical theatre 59–60; *liber corporum* 67
Schaffer, S. 138
Schiebinger, L. 91
Schrödinger, E. 124
sciences of 'Man' 40–1
scientific knowledge 27–8, 29–31
scientific modernity 43–4
scopes 90–1, 92
scribal culture 61–2
Serres, M. 130, 131
Shapin, S. 138
Shelley, M. 148–9
Shildrick, M. 41
Siegel, C. 106, 107
silver chalice 28
simulacra 134
simulation 81, 102; surgical 84–6
Smithsonian Institute 2
Sobchak, V. 130
Sochurek, H. 74
space 35–6, 63–78; actual and virtual 158–9; body as laminar terrain 63–6; body as readable terrain 66–70; tomographic 75–8, 95–6; virtual 4–5, 158–9; volumetricity 71–2
speculation 99
spirit 153–4
Spitzer, V. 12, 13, 26, 117, 167, 168
Stafford, B. 66, 168
standing-reserve 29–30
Starr, P. 92
Stengers, I. 32
Stix, G. 111, 168
Stone, A.R. 153
subject/object relations 41–2, 147
supernatural 152, 153–4
surface: and depth 159–60
surgery 25, 69; minimally invasive 88; open 88, 92–3; telematic 88–9; virtual *see* virtual surgery
surgical planning 86–7
surgical rehearsal 75, 87
surrogacy 72, 78, 82; dissection rehearsal and surgical rehearsal 74–5; and morphology 98–101

technē 28–9
technogenesis 157–62
technology 19–20, 24–50; distinction between biological and technological

161; human archives 37–42; posthuman medicine 42–50; technological frame 27–33; virtual *gestell* 24–7; visible human technics 33–7
telematic surgery 88–9; *see also* virtual surgery
telematicity 35–6
Terminal Choice 148
terrain 16; laminar 63–6; readable 66–70
Terranova, T. 43
text 66–70
theatre, anatomical 59–60, 71, 158, 166
three-dimensional imaging 9–12
Tiede, U. 35
Tierney, J. 141
time 79–80, 129–31
Tomas, D. 124, 150, 151–2
tomographic imaging; digital inscription 75–8; exscription and virtual surgery 94–8; *see also* computed tomography (CT) scans, magnetic resonance imaging (MRI)
trace, anatomical 58, 70, 78
transformations in technologic 49
transgenic techniques 25–6, 44
translation 43–4, 45
travelling the body 16; *see also* flythroughs
Treichler, P. 138
Tsiaras, A. 3
tutelage 67–8

ultrasound video 119
uncanny, digital 149–56
University of Michigan Scientific Visualisation Laboratory 17
University of Pennsylvania Medical Center 17

Van der Ploeg, I. 164, 168
Vasseleu, C. 114, 126, 133, 168
Vesalius, Andreas 64–6, 66, 69–70, 117
violence 55, 58; *see also* sacrifice
virtual colonoscopy 102–3

virtual endoscopy 73, 83–4, 102–6, 107
virtual *gestell* 24–7
virtual map 70–5
virtual modelling 25
virtual reality 5
virtual space 4–5; and actual space 158–9
virtual surgery 21, 81–109; flythroughs 101–6; morphology and the anatomical norm 106–8; operative images 108–9; radiographic exscription 91–3; surrogacy and morphology 98–101; televising interior morphologies 89–91; tomographic exscription 94–8
Visible Human Project (VHP) 1–23, 137–8; brief history 8–18; as human archive 37–42; uses of the data 16–18, 26
Visible Male 13–15, 17; *see also* Jernigan, Joseph Paul
Visible Woman 15, 17–18, 54, 56, 111
visual ordering 30–1
vitality *see* life
Vitruvius Pollio, Marcus 59, 165
volume 9–12; virtual map and surrogacy 71–2

Waldby, C. 18, 39, 165, 168
Waldorp, M. 38, 128, 168
Ward, B. 168
Wark, M. 4
Weber, S. 28, 159, 163; differential specificity 59; emplacement 31, 41; technics 29, 32, 41, 42, 45
Weiner, N. 110, 124
Wells, H.G. 13, 138
Wilkie, T. 26
Wirthin, K. 128, 168
Wolfe, C. 42, 52, 53
working objects 98–101

X-Files, The 148
x-rays 89, 91–3, 94–5, 167

Young, K. 93, 140, 146, 147